AN ECONOMIST'S GUIDE TO ENVIRONMENTALISM

AN ECONOMIST'S GUIDE TO ENVIRONMENTALISM

A Toolkit for Understanding and Solving Ecological Problems

JORDAN K. LOFTHOUSE

BLOOMSBURY ACADEMIC
NEW YORK • LONDON • OXFORD • NEW DELHI • SYDNEY

BLOOMSBURY ACADEMIC

Bloomsbury Publishing Inc, 1359 Broadway, New York, NY 10018, USA
Bloomsbury Publishing Plc, 50 Bedford Square, London, WC1B 3DP, UK
Bloomsbury Publishing Ireland, 29 Earlsfort Terrace, Dublin 2, D02 AY28, Ireland

BLOOMSBURY, BLOOMSBURY ACADEMIC and the Diana logo are
trademarks of Bloomsbury Publishing Plc

First published in the United States of America 2026

Copyright © Bloomsbury Publishing, 2026

Cover image: *Herd of Bison in Yellowstone* © stellalevi / Royalty-free / Getty Images

All rights reserved. No part of this publication may be: i) reproduced or transmitted in any form, electronic or mechanical, including photocopying, recording or by means of any information storage or retrieval system without prior permission in writing from the publishers; or ii) used or reproduced in any way for the training, development or operation of artificial intelligence (AI) technologies, including generative AI technologies. The rights holders expressly reserve this publication from the text and data mining exception as per Article 4(3) of the Digital Single Market Directive (EU) 2019/790.

Bloomsbury Publishing Inc does not have any control over, or responsibility for, any third-party websites referred to or in this book. All internet addresses given in this book were correct at the time of going to press. The author and publisher regret any inconvenience caused if addresses have changed or sites have ceased to exist, but can accept no responsibility for any such changes.

Library of Congress Cataloging-in-Publication Data

A catalog record for this book is available from the Library of Congress

ISBN: HB: 978-1-5381-8985-6
ePDF: 979-8-7651-5848-7
eBook: 978-1-5381-8986-3

Typeset by Integra Software Services Pvt. Ltd.
Printed and bound in the United States of America

For product safety related questions contact productsafety@bloomsbury.com.

To find out more about our authors and books visit www.bloomsbury.com
and sign up for our newsletters.

This book is dedicated to my father Kory Lofthouse and my mother Debbie Lofthouse, who taught me to work hard, think critically, and appreciate the wonders of nature.

CONTENTS

Acknowledgments ix

Introduction: Economics and Environmentalism 1

PART ONE An Economic Toolbox 13

1 Rational Choice Theory 15

2 Tradeoffs and Opportunity Costs 25

3 Marginalism 31

4 Institutions 37

5 Property Rights 43

6 The Entrepreneurial Market Process 51

7 Supply and Demand 61

8 Unintended Consequences 69

9 Public Choice Theory 75

10 Polycentric Governance 89

PART TWO Applying Economic Tools to Environmental Problems 99

11 Positive-Sum versus Negative-Sum Environmentalism 101

12 Species Conservation 131

13 Land Conservation 149

14 Water Scarcity and Water Markets 161

15 Climate Change 175

Conclusion 191

Notes 198
Select Bibliography 221
Index 231
About the Author 236

ACKNOWLEDGMENTS

This book would not have been possible without the guidance, support, and encouragement of many individuals. First, I want to thank Christopher Coyne, who provided helpful feedback on the full manuscript. I am also grateful to all my colleagues in the F. A. Hayek Program for Advanced Study in Philosophy, Politics, and Economics, including invaluable insights from Paul Dragos Aligica, Peter Boettke, Rosolino Candela, Ginny Choi, Kristen Collins, Erwin Dekker, Nathan Goodman, Stefanie Haeffele, Bobbi Herzberg, Arielle John, Jayme Lemke, Mikayla Novak, Solomon Stein, and Virgil Storr. I would also like to thank Deni Remsberg, Associate Managing Editor at Bloomsbury, for her helpful guidance through the publication process. Last, I am profoundly thankful to my husband Troy Olsen for his unwavering support and encouragement. He has taught me to be a better teacher and communicator.

Introduction

Economics and Environmentalism

Drastic times call for drastic measures—a belief that many environmental activists have taken to heart. In recent years, new forms of high-profile and disruptive protests have brought attention to dire ecological problems. For example, in 2022, activists from Last Generation blocked the runways at the Berlin and Munich airports, calling on the German government to stop subsidizing air travel.[1] The group's members have also glued themselves to busy commuter roads to demonstrate against what they see as the German government's failure to act against climate change. One of Last Generation's members said, "It's absolutely crazy to stick yourself to the road with superglue," but she retorted, "If we wanted people to like us then we'd do something else but we've tried everything else. We've asked nicely. We've demonstrated calmly."[2]

Also in 2022, members of the group Just Stop Oil glued themselves to a Vincent van Gogh painting in London. One of the activists who glued his hand to the frame, said, "Sorry everybody, we don't want to be doing this. We're here glued to this painting, this beautiful painting, because we're terrified for our future."[3]

In 2021, five activists staged a hunger strike in front of the White House, demanding more action from the Biden administration on climate change. In a letter, the strikers wrote, "We will continue to sit starving outside the White

House everyday until you [President Biden] use your power as elected president of the United States to deliver your mandate for bold, and transformative climate action with justice and for jobs."[4] One of the participants said, "Hunger symbolizes what we're starving for—what we deeply need in this moment."[5]

While not all people who are environmentally conscious are willing to glue their bodies to roads or engage in multiday hunger strikes, tens of millions of people across the globe are sympathetic to ecological problems. For instance, a large majority of Americans say that they want the federal government to enforce environmental regulations more strongly, set higher emissions standards for automobiles, spend more government money on developing solar and wind power, and impose mandatory controls on carbon dioxide emissions and other greenhouse gases.[6]

Environmentalists care deeply about a broad range of ecological issues. They fear the extinction of countless species. They protest the destruction of unique and precious landscapes. They see life-sustaining water becoming increasingly scarce in many places, threatening natural systems and human societies. They experience the hardships of climate change—one of the largest and most important challenges that humanity has ever faced. These environmental problems, and many more like them, leave us with two important but difficult questions: What are their root causes, and what can we do to address them?

Scholars in a variety of academic fields have potential answers for these two questions, but the field of economics has particularly useful insights. This book uses concepts and theories from economics to analyze why environmental problems emerge and how we might be able to address them. For people who don't study economics, they might have misconceptions about what it actually is. A popular stereotype of economists is that they predict the stock market or focus solely on money, using mathematical equations or complex graphs to make people richer at the expense of other important values. Perhaps a negative view of economists is that they are cold, detached, and overly abstract, neglecting the social or moral implications of real-world issues.

In reality, economics is a much broader field of study and includes a wider range of insights than just doing mathematics or talking about making money. However, even economists can't quite agree on what exactly "economics" is. Several famous economists have given their own definitions. Jacob Viner gave a tongue-in-cheek answer, saying, "Economics is what economists do,"[7] but other economists have attempted to describe their field a bit more specifically. John Maynard Keynes called economics "a method rather than a doctrine, an apparatus of mind, a technique of thinking, which helps its possessor to draw correct conclusions."[8] Lionel Robbins called economics the study of "human behaviour as a relationship between ends and scarce means which have alternative uses."[9] James M. Buchanan said that "Economics is the study of the whole system of exchange relationships," or in other words, it's the study of exchange and the social systems in which exchange takes place.[10]

Friedrich A. Hayek preferred the term "catallaxy" rather than "economics" to describe what economists study. The term "economy" comes from the Greek work *oikos*, meaning "household," implying a sense of planning and supervision over how resources will be used. In contrast, the Greek verb *katallattein* means "exchange," with connotations of "admitting into the community" or "changing from an enemy to a friend." As Hayek argued, economics is about people exchanging with one another, as opposed to a planner allocating resources for an entire society. In Hayek's view, the term "catallaxy" better describes the "order brought about by the mutual adjustment of many individual economies in a market."[11]

By combining these definitions, economics is simply a way of thinking about the social world, using a particular set of assumptions and tools. The main focus of economics is on human choice and how humans interact and exchange with one another. Adam Smith, who lived in Scotland in the eighteenth century, has been called the Father of Modern Economics because he developed the tools to give a systematic explanation of economic outcomes. However, Smith didn't necessarily think of himself as an economist. He was a moral philosopher, and he was curious why some places are rich and other places are poor. In 1776, he published his most famous book *An Inquiry into the Nature and Causes of the Wealth of Nations* (usually just shortened to *The Wealth of Nations*) to answer this fundamental question. He described how countries become rich when they have stable social rules that allow people to specialize their labor and trade with one another.

However, Adam Smith only provided the starting point. In the nearly 250 years since Smith wrote *The Wealth of Nations*, many other brilliant minds have built on his foundation and have refined his insights. We now better understand human choice, how markets work, the role that governments can play in helping or hindering economic growth, and how social relations and cultural beliefs affect economic processes.

In the mid-nineteenth century, the Scottish writer and essayist Thomas Carlyle coined the term the "dismal science" to refer to economics.[12] Despite the gloomy moniker, economics is especially helpful for environmentalists because it can explain why environmental problems arise, and it also provides a framework to evaluate the vast array of potential solutions. Economics isn't dismal at all because it provides insights that should give environmentalists hope to address environmental problems.

This book has two purposes. The first is to introduce readers to foundational economic concepts, which serve as "tools" to understand the causes of environmental problems and to evaluate potential solutions. The second purpose is to apply these economic tools to real-world issues, like species extinction, land degradation, water scarcity, and climate change. Many more environmental problems exist in the world, so the hope is that readers will take the tools and apply them more widely to other issues.

To assemble the toolkit, this book summarizes and builds on ideas from a variety of scholars in economics or closely related fields, including several Nobel Laureates like Friedrich A. Hayek, James M. Buchanan, Elinor Ostrom, Ronald Coase, and Douglass North, as well as prominent scholars like Adam

Smith, Gordon Tullock, Israel Kirzner, Joseph Schumpeter, Julian Simon, and William A. Niskanen. It also builds on the pioneering work done in "free market environmentalism" by Terry L. Anderson and Donald R. Leal.[13] Although these scholars represent many viewpoints and approaches, the ideas highlighted in this book reflect principles that have (relatively) broad acceptance within the field of economics. It's important to note, however, that economists are diverse, and they often disagree on assumptions, methodologies, and policy implications. This book presents one view of economics, but other views exist. Some economists might say that certain ideas or thinkers in this book have been overemphasized, while others were underdeveloped or neglected. Of course, this book is not a definitive proclamation or the final word. Instead, it's an invitation to explore how ideas from economics can be useful in understanding and addressing environmental issues.

The first part of this book, consisting of chapters 1 through 10, assembles a toolbox of ten foundational concepts from economics. Although economists often use math and graphs to communicate their ideas, the underlying way of thinking is more important than any particular method or technique that economists use. Below is a very brief overview of the tools:

1 Economists' most basic assumption is that all people act purposively, meaning that they try to reach their goals while also responding to their incentives and constraints. In other words, they are *rationally self-interested*. Self-interested does not necessarily mean selfish, and rational does not mean infallible. People strive to achieve their goals, even though they have imperfect knowledge and their circumstances are not ideal.

2 Scarcity is an unavoidable fact of life, meaning that resources are limited but human desires are unlimited. Since that is the case, every choice—whether as an individual or as a group—involves *tradeoffs and opportunity costs*, meaning that something is always given up when

one course is chosen over another. This insight implies that there aren't true "solutions" to problems, only tradeoffs. In other words, addressing a problem requires a compromise because some other desirable alternative is necessarily foregone.

3 People *think on the margin*, or in other words, people make decisions based on small, incremental changes rather than big, all-or-nothing choices.

4 All human action takes place in *institutions*, which are the formal or informal "rules of the game" that shape people's incentives and constraints.

5 *Property rights* are a type of institution that specifies who has the decision-making power over how particular resources are used. Most environmental issues arise when property rights are not clearly defined or not enforced well.

6 The *market is a process driven by entrepreneurial action*, which allows people with different plans to coordinate with one another. Markets are incredibly powerful because they facilitate the discovery, aggregation, and communication of knowledge that allows people to make their lives better.

7 The logic of *supply and demand* is a simplified but useful representation of how markets work, which helps us evaluate what markets do to coordinate between buyers and sellers and what happens when policymakers intervene.

8 All human action is followed by *unintended consequences*, but some of those consequences are socially helpful while others are harmful.

9 When evaluating public policies, such as legislation or regulations, we need to know the incentives and constraints that government officials face, which is a subfield of economics called *public choice theory*. Public

policies are the result of many individual politicians and bureaucrats responding to their individual self-interest, which may sometimes align with the broader public's interest. However, sometimes public policies concentrate benefits on well-connected special interest groups at the expense of other groups, like taxpayers at large.

10 The real world is complex, and different institutions have different characteristics. To cope with the complexity, we often need to use a *polycentric institutional arrangement*, which refers to a social structure in which decision-making is made in many different locations and at many different levels. The decision-makers are independent in some ways and interdependent in others, as opposed to a centralized, top-down approach.

Part two of the book contains five chapters that apply these economic tools to specific, real-world environmental issues. Chapter 11 uses these tools to understand the "big picture" of environmental problems by showing their general causes and examining some potential solutions. The chapter focuses on institutions, entrepreneurship, and positive-sum interactions. First, institutions channel and shape human behavior, particularly the institution of private property rights. Most environmental problems arise when private property rights are not clearly defined or enforced well. For example, you have property rights to your backyard, and people have a strong incentive not to throw their garbage there, otherwise they would face legal action or social shaming, or both. However, it's nearly impossible to assign property rights to the air, so it's much easier to spew pollution into the air since no one owns it. That's when we turn to government to regulate these spillover effects, but not all regulations are created equal.

This is where entrepreneurs come in. In markets, entrepreneurs can and do discover innovative ways to produce greener products. In government, entrepreneurs can and do find more effective public policies to address environmental issues. In civil society, entrepreneurs can and do create new clubs, associations, and nonprofits that shape the way people think and act, especially as it relates to environmental issues.

However, particular forms of institutions and entrepreneurship are not always beneficial. That's why it is important to look for institutions that channel entrepreneurship in positive-sum ways, as opposed to zero-sum or negative-sum ways. In positive-sum interactions, all people are made better off. Voluntary market exchanges are positive sum because people only exchange when both sides determine that trading will improve their current situation. In zero-sum interactions, one side wins while the other side loses. Worst of all, negative-sum interactions mean that society overall is worse off because the losers lose more than the winners win. In many cases, politics is a zero-sum game that can devolve into negative-sum fights because one group can use the coercive power of government to take from one group to give to another.

Environmental problems are more likely to be solved when the solutions are positive sum, and the "free market environmentalism" approach focuses on positive-sum solutions. In contrast, political environmentalism is the traditional way of doing natural resource economics that focuses on using government regulations to prevent environmental disaster.[14] Anderson and Leal—founders of the free market environmentalism approach—caution against falling into the "nirvana fallacy," which involves "comparing a less than perfect system with one that is assumed to be so."[15] Using private property rights and markets may not be able to solve every environmental problem in the world, but neither can the traditional government approach, so the trick is looking for institutions that align private incentives with desirable environmental outcomes.

Chapter 12 examines the threat of extinction to different species and success stories of how some species have been successfully conserved. One example looks at the conservation of African rhinos. In most African countries, it has been illegal for individuals to own wild game. Many governments tried to increase rhino populations and stop poaching, but populations continued to trend downward. Some countries, such as South Africa, Namibia, and Zimbabwe, have changed their laws in recent years to allow for private individuals to own rhinos. The new establishment of private property rights changed the incentives that people faced, and the new owners of rhinos began to breed them, increasing the populations. Today, South Africa has the world's largest population of privately owned rhinos, which are used for conservation purposes, ecotourism ventures, and trophy hunting.

Chapter 12 also looks at the conservation of the greater sage-grouse in the western United States. The sage-grouse is a ground-dwelling bird that lives on the "Sagebrush Sea" that extends between the Rocky Mountains on the east and the Sierra Nevada/Cascade Mountains on the west. In 2010, the US Fish and Wildlife Service announced that it was thinking of listing the greater sage-grouse as "threatened" or "endangered" under the Endangered Species Act, which would entitle it to many strict protections. However, these legal protections have huge impacts on private landowners and the users of public lands.

To avoid the ramifications of listing the bird under the Endangered Species Act, the eleven state governments with sage-grouse populations each tried different approaches to boost populations and protect habitat. Many of these approaches involved a variety of stakeholders, including federal, state, and local policymakers, as well as environmental groups, nonprofit organizations, scholars, and associations of for-profit companies. The collaboration among all these various people boosted the population of the greater sage-grouse, and the US Fish and Wildlife Service chose not to list the species under the Endangered Species Act, avoiding all the potential restrictions that would have come with the legal designation. The story of the sage-grouse shows how collaborative approaches and more on-the-ground decision-making can lead to effective environmental outcomes.

Chapter 13 examines different ways of protecting and conserving land. In many cases, some of the largest areas of protected land are under direct government control, such as national parks, national forests, and officially designated wilderness areas. However, government is not the only means to protect land. Private property owners play a vital role in ensuring that landscapes are conserved and protected. This chapter highlights an innovative approach of landscape conservation in Montana at the American Prairie Reserve (APR), which is a nonprofit organization that is trying to stitch together what is essentially a privately owned national park. In other words, APR receives donations, and it engages in "willing buyer, willing seller" transactions with landowners in the region. The end goal is to amass enough privately owned land and adjacent federal/state grazing land so that this area of the Great Plains is restored to its pre-colonial setting, with all kinds of wildlife roaming freely.

Of course, this project has faced challenges, but it has tried to meet those challenges with an entrepreneurial spirit. Since growing numbers of wildlife could have negative spillover effects on neighboring ranchers and farmers, APR has started incentive programs that pay local ranchers and farmers to engage in wildlife-friendly practices, changing the presence of wildlife from a detriment to a benefit. However, other groups don't agree with APR's project, and they have sought to stop it from moving forward by persuading policymakers to change the rules about how leased federal grazing land can be used. The story of APR shows how private property rights can facilitate large-scale land conservation, but things become more complicated and conflict-ridden when the politics of public land get involved.

Chapter 14 explores water conservation and the various means we have for allocating this precious but scarce resource. Water markets can be and have been a very important way of allocating water efficiently. However, sometimes water markets don't function well because property rights are insecure, or public policies create perverse incentives. Australia's water-market system, while not perfect, has been largely successful and effective at managing the scarcity of water on a very arid continent. One of the largest benefits of Australia's water-market system is that market prices and the trade of water shares allows for flexibility when water is abundant or scarce. Droughts are hard to predict, and water markets allow people to adjust how they use water when a drought occurs. Other places across the globe have begun embracing water markets, which has reduced the degree of conflict over water.

As opposed to markets, the political allocation of water has not always been successful, and it often sparks conflict. In many countries—such as Bangladesh, Tajikistan, Malaysia, Yugoslavia, Angola, East Timor, Namibia, Botswana, Zambia, Ecuador, and Peru—the political allocation of water has either directly or indirectly led to armed conflict and made it more difficult to adjust to changing conditions. In a world where water is becoming scarcer in many places, water markets may be the best arrangement for dealing with the uncertainty of the future. Water markets also have the additional benefit of providing a more peaceful way of resolving disputes because people can bargain with one another, rather than fight.

Chapter 15 looks at air, especially the emission of pollution and greenhouse gases. Human-caused climate change is an incredibly complex issue because there are many sources of greenhouse gas emissions and there are also many effects. The causes of climate change involve transportation, agriculture, and industry, each of which are necessary for modern life. The effects of climate change differ from place to place. Some locations are getting drier, while others are getting wetter. Sea levels will continue to rise, threatening populations along the coasts. The oceans will continue to heat up and acidify, harming ecosystems, especially coral reefs. To make matters more complicated, different people have different capabilities in dealing with the localized effects of climate change.

In the face of all this complexity, Chapter 15 argues that addressing climate change—both in terms of mitigation and adaptation—is best done through a polycentric approach. The collaboration and competition of many companies in markets often leads to products that are better for the environment. Different levels of government can experiment with different policies in different places, and they can learn from one another. Many different clubs, associations, and nonprofits can provide checks and balances on market firms and government policymakers to ensure that climate action takes place.

The alternative to a polycentric approach is a monocentric one that is top-down and centralized, which can have many shortcomings. When there's just one approach, there are fewer opportunities for experimentation and learning. A single approach means that any failure affects the whole system. If the people under that one approach don't agree with it, they may choose to ignore or sabotage the policies. A polycentric approach is a better way to address the many problems associated with human-caused climate change because it leads to better knowledge, more resilience, and more buy-in from people on the ground. In short, effective mitigation and adaptation for climate change will involve a large degree of action from the bottom up, and we cannot simply rely on policies from the top down.

Chapter 16 concludes the book with some bigger considerations. It's not an understatement that people disagree on environmental issues, like how serious problems really are or how to address them. These disagreements can lead to harsh words, and harsh words can lead to violence if we are not careful. Violent conflict can undermine other ideals that we might cherish, like peace, liberty, and democracy. We can find ways to address environmental problems and also preserve other values that we hold dearly. By using the tools discussed in this book, anyone can act entrepreneurially and discover solutions to environmental problems that are win-win.

PART ONE

AN ECONOMIC TOOLBOX

1

Rational Choice Theory

All human beings use tools to help them accomplish their goals, and scientists use particular tools to help them understand the world better. Biologists use microscopes to observe single-cell organisms, helping us develop an understanding of the causes of disease. Astronomers use telescopes to observe faraway stars and galaxies, allowing us to better comprehend the cosmos. Medical doctors use magnetic resonance imaging (MRI) machines for seeing inside the body to diagnose diseases so that we can treat human suffering.

However, not all tools are physical objects like microscopes, telescopes, or MRI machines. In the social sciences, including economics, the tools are a set of assumptions, frameworks, theories, models, methodologies, and techniques that help us to better understand the causes and consequences of human action. When Adam Smith—the famous eighteenth-century moral philosopher and economist—questioned why some places are rich and some places are poor, the tools he used to answer his question were in his mind. Economists today still use the intangible tools that Adam Smith and his intellectual heirs developed, but they pair those intellectual tools with modern technology to improve their understanding of the world.

By assembling a toolbox of economic concepts, we can begin to answer many important questions in social science: What causes some places to become wealthy? How does higher education affect people's livelihoods? Why does inflation happen? What are the effects of subsidies on farm production and food prices? What causes housing to become unaffordable? Why has the War on Drugs largely failed? Economic tools are applicable to nearly every aspect of human life, and economists have used these tools to produce compelling and useful answers to some of the most pressing questions in society. Environmental issues are no different.

Before we start putting together our economic toolbox, we need to distinguish between positive and normative analysis. Positive analysis is rooted in statements of fact or observation, based in theory. In other words, positive analysis is based on "is" or "if-then" statements. For example, in physics, if you drop a ball in a vacuum near the surface of the earth, then it will accelerate at approximately 9.8 m/s^2. The acceleration of gravity on Earth is simply a fact that we have observed through measurement. However, in contrast to positive analysis, normative analysis is more philosophical because it is rooted in moral or ethical considerations. In simpler terms, normative analysis is based on "should" or "ought" statements. For example, saying that a society should have political equality or economic equality is normative. It's about determining what is "good" versus "bad."

Economics is a *positive* social science with *normative* implications. As with all social issues, people have many strong normative feelings, but economics itself cannot tell us what is good or what we should do. We need philosophy to do that. We can start with a normative, philosophical stance, such as "Protecting species from extinction is a good thing." Once we have a normative goal, we can pivot to the positive analysis of economic thinking. *If* our goal is to protect species from extinction, *then* what will be the most effective or efficient means to attain that goal? We can use the tools of economics to answer these positive questions. However, we might not like the answers we get. We can pivot back to normative analysis to evaluate the outcomes of our positive analysis.

Another example might be helpful. Let's say we have the normative goal of making electric cars more affordable to reduce greenhouse gas emissions. We might think that cheaper electric cars and lower emissions are good things. Once we have our normative goal, we can pivot to positive analysis to figure out the most effective or efficient means to achieve it. We can say that we want to create a government policy capping the price of electric cars to ensure that less wealthy individuals can afford them. However, once we start using the positive tools of economics, we can see that price controls may not produce the results we want. Using the logic of supply and demand (discussed more in Chapter 7), we can see that a price control policy, like a price cap, will create a shortage of electric cars. We can also see that price controls have other unintended consequences (discussed more in Chapter 8), such as lower product quality, more under-the-table payments or "black markets," less investment in innovation-sparking research and development, and increasing market concentration in larger firms. Once we have used positive analysis to determine what the outcomes will be, we can pivot back to normative analysis to determine whether those outcomes are good. We might conclude that the outcomes from price controls are worth it, or we might decide to try another means of improving the affordability of electric cars without the undesirable outcomes.

The economic tools that we are putting into our toolbox are positive, but they have important normative implications. This is why economics and philosophy often go hand in hand. We need a philosophical framework to figure out what good goals are and whether the outcomes of an action are good. However, we need an economic framework to do the cold, hard analysis of *how* and *why* things work. In this book, the normative ideas are left up to the reader. The goal here is to present a set of positive tools, and the reader can decide how to use them. A hammer and a saw are tools for building a house, but those tools can't tell you what a beautiful or an ugly house looks like. With that preface, let's proceed with putting the first tool in our economic toolbox.

The foundation of economics is rational choice theory, which assumes that all people—regardless of their location, religion, culture, or socioeconomic status—are rational. The term "rationality" has many connotations, but economists generally use a broad, simple definition: People pursue their goals to the best of their ability, taking into account their limitations and the limitations of their environment.

For example, an archer has the goal of using his bow and arrow to hit the bull's eye on a target. He rationally tries to shoot the bull's eye. Even if he isn't very good at archery, the fact that he places an arrow on the bowstring, pulls the bowstring back, and attempts to aim at the target means that he is rationally trying to accomplish his goal. The archer may miss the target, but no one would say that he is irrational. However, an irrational archer would face in the opposite direction from the target and still expect to hit it. In other words, rationality means that people purposefully choose means that they believe will help them accomplish their ends. Rationality does not mean "free from mistakes." Performing a rain dance is a rational act if you genuinely believe that a rain dance summons a storm. Rationality is simply congruence between actions and goals.

This is related to Nobel Laureate Friedrich A. Hayek's idea that the facts of the social sciences are what people think and believe.[1] Hayek made a distinction between the natural sciences and the social sciences. The natural sciences generally focus on objective, measurable phenomena, like chemistry, physics, and geology. However, the social sciences, like economics, sociology, political science, and psychology, focus on understanding human behavior, which is rooted in individuals' perceptions, beliefs, and intentions. Economics is different from physics because atomic particles don't think and feel and talk back to the physicist. An economist can talk to another person about their subjective preferences and intentions, allowing us to better understand and predict behavior.

Rational choice theory might sound a bit unrealistic at first because people clearly do things that actively harm their interests. Also, they sometimes have mental issues that cause them to choose means that are self-destructive or means that aren't appropriate for achieving their ends. Discovering why people do irrational things is the realm of psychology rather than economics. Psychology is an important science, but most economists generally assume that people act rationally for the purposes of doing analysis. In other words, most economists start with the assumption that people act rationally, and any apparent deviation from rational behavior must be explained by some other factor.

However, one important thing to keep in mind is that behavior that seems irrational at first is often very rational when you understand the full story. For example, millions of people vote every year, even though their vote is very unlikely to affect the outcome of an election. If we assume that people are rational, then we would assume that most people wouldn't vote because of the math. Just imagine the last election you voted in. How close was the election? Did the deciding factor come down to exactly one vote? It's very unlikely that an election, especially a nationwide one, comes down to one vote. You could have stayed home, and your vote wouldn't have changed the outcome of the election. Of course, this logic holds only when the size of the group that is voting is fairly large. If you're on a committee of only five people, then your vote is important because there is a very high chance that your vote will be the deciding one. However, when millions of people vote, the likelihood that your vote matters is essentially zero. In almost every case, any individual could have not voted, and the results of the election would have remained the same.

If that is the case, then people must be irrational when they vote. However, a broader perspective explains this apparent irrationality. People might vote as a form of entertainment, self-expression, or fulfilling a sense of civic duty.[2] There are many more examples of seemingly irrational behavior that are actually perfectly rational when you understand all the details of the situation, such as the workings of pirate ships, medieval maledictions, and wife sales in Industrial Revolution–era England.[3] Practices and behaviors that make people worse off are unlikely to survive for a long time, so most long-lasting practices are usually rooted in an unseen or obscured form of rationality.

Two of the most important components of rational choice theory are the concepts of incentives and constraints. In the simplest terms, an incentive is something that motivates or encourages a person to act in a certain way. Another way to frame incentives is in terms of costs and benefits. Whenever any action has a relatively large degree of benefit to a relatively low level of cost, there is a strong incentive to engage in that action, and the reverse is true. For most people, they have a strong incentive to go to work on days when they would rather stay home, but why? Not showing up to work would probably get you fired, which would take away your income, which is a huge cost compared to the relatively small benefit of staying home and relaxing.

One of the clearest examples of incentives in action was during the early settlement of Australia. Originally, the British Empire that conquered Australia used it as a prison colony for fairly low-level criminals, like petty thieves, political prisoners, debt defaulters, vagrants, and prostitutes. During the late eighteenth and early nineteenth centuries, when the British government started shipping prisoners to Australia, they would contract with private shipowners to carry the prisoners to the opposite side of the world. The journey would take months, and many convicts would die along the way. The ships were overcrowded, food was scarce, and medical treatment was basically nonexistent.

British officials were alarmed at the number of convicts that were dying on the way to Australia, so the government decided to pay the captains a bonus for each convict that made it alive. With this new payment scheme, the captains had a new incentive structure. It was in each captain's personal interest to make sure that as few convicts died on the journey as possible. Before, the convicts weren't worth much to the captains, whether they were dead or alive. By changing the way that captains were funded, the British government made living convicts much more valuable than dead ones, and the captains responded rationally to their incentives by providing better services to the convicts.[4]

A constraint is a limit or restriction that affects what somebody can choose or do. One of the clearest constraints that everyone understands is a budget constraint. Imagine you want to buy an oceanside mansion in Malibu, California. Unless you have several million dollars in cash, or you have connections with people willing to extend a long line of credit, your income is a constraint on your ability to live your Malibu dream. The laws of physics are also a constraint—even if you want to fly, you can't flap your arms fast enough to provide the necessary lift to get off the ground. Constitutional rules are constraints that forbid certain kinds of actions by government officials. In the United States, for example, government officials are constrained in limiting your freedom of speech, except in some rare and extreme circumstances.

Some constraints are softer than others. Constitutional constraints on government officials are bent all the time, and in certain places, the constraints are very loose. On the other hand, the laws of physics are relatively hard constraints for nearly all actions. However, it's important to acknowledge that seemingly insurmountable constraints can change over time. For instance, in October 1903, the *New York Times* published an article that predicted that a functional flying machine would take between one and ten million years to achieve, but the Wright Brothers took their first flight only two months after that article was published. Then in 1910, famous astronomer William H. Pickering expressed his skepticism of air travel, saying, "The expense would be prohibitive to any but the capitalist who could use his own yacht," but the first commercial passenger flights began in 1914.[5] By the 1950s, the upper middle class could afford cross-country or intercontinental flights, and by the 1970s, millions of middle-class Americans could afford to travel by plane almost anywhere.

Related to rational choice theory is the concept of "methodological individualism"—a phrase with a lot of syllables for a relatively simple idea. It essentially means that only individuals act and choose. In other words, the individual is the basic unit of analysis. Groups can obviously come to decisions, but those decisions are the product of the discussions and interactions of the individual members of the groups.

We often talk about groups or collective entities as if they choose, but that can hide the actual interesting part of how the world works. For example, if you watch the US news, a common phrase is "Congress passed a bill." While it may be technically true that Congress passed a bill, Congress is not just one thing. Congress is made up of the House of Representatives, which has 435 voting members, and the Senate, which has 100 people. To pass a bill, both chambers must agree on the same wording of a proposed law. In the House, the 435 members try to persuade one another and make deals with one another to accomplish their goals. In the Senate, people do the same thing. If there are differences between the two versions of the same bill that each chamber passed, then a special "conference committee" is formed from a few members from both the House and the Senate. In this committee, the participants try to iron out the differences and produce a final version of the bill called a "conference report," which goes back to the full membership of both chambers. Then the members in the House and Senate give one final yes or no.

The civics lesson above is to demonstrate that saying "Congress passed a bill" is a very over-simplistic view of the world and does not help much with analysis. To really understand the social world, we need to look at the individuals who make up a group. We must ask ourselves many questions about those individuals: What are each of their unique incentives? What constraints do they face? How do they interact and exchange with one another? By using methodological individualism as a tool for analysis, we can get to the bottom of how group decisions and collective action actually happens.

We see rational self-interest play out in environmental policymaking all the time. Politicians from areas with large oil or coal industries try to ensure their reelection by protecting jobs in their districts and maintaining financial contributions from energy companies. Other politicians might want to subsidize solar or wind energy projects in their home districts, knowing the government funding will benefit local voters. Politicians might introduce or support popular environmental policies, like banning single-use plastics or expanding renewable energy incentives, just before an election to gain favorable media coverage and voter support. Of course, the politicians in these examples are human, and they may have some altruistic motivations for their choices. But, like all people, they are also rationally self-interested, which means that they strategically frame their rhetoric and actions for their benefit. In many cases, policies that benefit politicians also benefit the general population, but that is not a guarantee.

2

Tradeoffs and Opportunity Costs

One of the foundational facts about the world is scarcity, which is the central issue of economic analysis. There is not enough of everything to fully satiate everyone's desires. In other words, humans have virtually limitless wants, but all resources are limited to some degree. The fact that there are more wants than available resources means that people must make choices about what to produce and how to produce it. Since the things we want are not found in unlimited quantities, this leads to conflicts, bargaining, politics, market interactions, and so on.

Due to the scarcity of all goods, individuals have to decide how much of one thing they want over another. That decision is a tradeoff, which is the sacrifice of giving something up for something else. Tradeoffs highlight the fact that people value things *subjectively*. In simpler terms, people are different, and the value of any action or item is determined by an individual's preferences and perceptions. Goods and services don't have any inherent, objective, or intrinsic value. The only value that something has is what people ascribe to it. Value is truly in the mind of the beholder.

A fingerpainting by a kindergartener and a modernist masterpiece by artist Jackson Pollock are essentially the same thing—colors splattered across a piece of paper. However, a child's fingerpainting will have a very low value on the market (if it can be sold at all), but a painting by Pollock is worth millions upon millions of dollars. The difference is that many people subjectively ascribe high value to Pollock's splatters and are willing to spend exorbitant amounts of money to own one.

Let's say that you only have enough cash to buy a bike or a snowboard, but not both. If you choose to buy the bike, you have traded off the snowboard. Or, let's say that you are deciding whether to take time off from your job to be with your family on a holiday or to work over the holiday for overtime pay. If you choose to spend time with your family, you are trading off the extra money you could have made.

Economics can't tell you whether the tradeoff you chose was good or bad (you'll need philosophy for that), but it's an undeniable fact that every choice has a tradeoff. A study of rational action and tradeoffs takes a person's subjective preferences as the starting point and analyzes a person's behavior from there. It's natural for everyone to make normative judgments about whether another person's preferences are "good" or "bad," but economic analysis doesn't need to get caught up in deciding between good and bad preferences.

Economists often say that problems don't really have "solutions," they only have tradeoffs. Sometimes the tradeoffs might be worth it (at least according to the chooser's subjective standards), but that doesn't mean a decision is costless. Every tradeoff has an opportunity cost, even if it is not a monetary cost. An opportunity cost is the value of the next best alternative that you give up when making a decision.

For example, if you choose to spend your time going to the gym, you can't spend that same time going to the library. If you choose to spend $100 going to a concert, you can't spend that same $100 to go to a restaurant. Since people have different subjective preferences, a tradeoff that one person is willing to make may be unthinkable to someone else with different preferences. Something is always given up.

Many opportunity costs are mundane, like choosing one breakfast cereal over another, but other opportunity costs can be very difficult to navigate. For example, in the Mojave Desert of southern California, tens of thousands of acres of land have been used for solar power farms. Of course, this renewable energy source sounds great, especially with the threats of climate change. However, all projects—even seemingly good ones—have opportunity costs. As the solar farms continue to expand across the desert, other desirable things are sacrificed. The area in which the solar farms now exist is rare habitat for a number of endangered species, such as the desert tortoise. The fragile ecosystem also contains carbon-capturing woodlands, as well as ancient Indigenous cultural sites. The people who currently live in the area have to deal with the side effects of all the new construction of the booming solar farms. One resident living near the solar farms said, "We're a senior community, and half of us now have breathing difficulties because of all the dust churned up by the construction. I moved here for the clean air, but some days I have to go outside wearing goggles. What was an oasis has become a little island in a dead solar sea."[1] Even something as desirable as green energy has many opportunity costs, and sometimes the people making the decisions don't bear the full opportunity costs of their decisions, which spill over onto other people.

There are many other examples of difficult environmental tradeoffs, each with their own opportunity costs. One is the tension between resource extraction and economic development. Activities like logging, mining, or agriculture are factors in economic growth, providing for people's livelihoods and delivering the goods and services that people desire to live fulfilling lives. However, resource extraction also can and often does lead to habitat destruction, carbon emissions, and loss of biodiversity. The opportunity cost of limiting economic growth for environmental benefits is all the foregone goods and services that people could have had. There is no easy or straightforward way to evaluate these tradeoffs, especially since different people have different preferences and ideas about morality.

Another example of environmental tradeoffs is the recent bans of single-use plastics across the globe. People generate millions of tons of plastic waste every year, but plastic is also important for making modern life easier and more convenient. When policymakers ban single-use plastics, producers of goods must find alternatives, like biodegradable products, which also have environmental and economic impacts. In many cases, the alternatives to plastic are paper and wood, which requires trees to be grown, harvested, transported, and processed. These processes also have environmental impacts, just like the creation of single-use plastics. In other words, banning single-use plastics can reduce pollution and environmental damage in some ways, but it can also lead to other environmental impacts. In essence, when we ban single-use plastics, we trade off one form of environmental damage for another. Overall, we don't yet have an accurate sense of the long-term implications of switching from plastic to other materials available for single-use items. More "life cycle assessments" are necessary to figure out how alternatives like wood and paper compare to plastic. It might be that in the long term, wood and paper for single-use items are better for the environment overall, but even so, the production of wood and paper will still have environmental impacts.[2]

Another example is the recent push to shift from conventional agriculture that uses synthetic pesticides and fertilizers to organic farming. Synthetic pesticides and fertilizers have environmental impacts and effects on human health, such as accidentally killing non-targeted species, disrupting native ecosystems, increasing toxic chemicals in food chains, and leading to "dead zones" in waterways and oceans.[3] Organic farming, on the other hand, can improve soil health and reduce chemical runoff. However, synthetic pesticides and fertilizers allow us to grow a larger amount of food on a smaller amount of land. Organic farming often has lower yields and greater yield variability.[4] As such, if the entire agricultural system were to switch to organic farming, we would need far more farmland than we have now, which might require encroachment into previously undisturbed ecosystems. Thus, the main tradeoffs of switching completely to organic farming are (1) more uncertainty regarding how much food we will have and (2) more land being used for agriculture.

It is important to note, however, that tradeoffs are not static. Tradeoffs depend on a combination of available resources, technologies, and preferences, all of which change over time. As new resources are discovered, or as new technologies emerge, or as people's preferences shift, the nature and magnitude of tradeoffs will also evolve. Tradeoffs that seem especially difficult in the short term might not be especially relevant in the long term.

3

Marginalism

In *The Wealth of Nations*, Adam Smith discusses a strange paradox that anyone can see. Water is essential for life. Diamonds are essentially just sparkly rocks, and they aren't necessary for life like water is. However, despite the necessity of water and the frivolity of diamonds, diamonds are much more valuable than water. Smith was undoubtedly a very intelligent person, but he couldn't figure out this paradox, and it would take almost a hundred years for people to solve it.

Three economists in the 1870s—Carl Menger, William Stanley Jevons, and Leon Walras—each independently discovered the solution to the "water-diamond paradox." Their discoveries completely changed economic theory, which is called the marginal theory of value or marginalism. Before this "Marginal Revolution," economists like Adam Smith and Karl Marx used the labor theory of value, which assumed that the value of a good or service comes from how much time and effort went into its production. The Marginal Revolution turned the labor theory of value on its head. Menger, Jevons, and Walras said that the value of goods and services lies in their marginal utility, as subjectively determined by consumers.

The answer to the water-diamond paradox is this: Individuals don't choose between all of the world's water versus all of the world's diamonds. If they had only those two choices, they would clearly choose water. However, individuals choose between *increments* or *margins* of a good. Based on their subjective preferences, each person decides whether it is worth having one more unit of water or one more unit of diamonds. For most people, water is very plentiful, and they don't have to think much about giving up or using another gallon of water. In the United States, most people who use municipal water pay only a fraction of a cent per gallon. However, diamonds are rare, and the average person owns very few of them.

Now imagine you've been stranded in a desert without water for a day, and you know that you'll certainly die if you don't get water soon. After stumbling around in the desert for a few more hours in the heat of the sun, you see an oasis in the distance. As you approach the oasis, you see that the oasis is fenced, and there is a guard standing at the gate. You ask him if you can enter and have some water. He answers that he will charge you $100 for a gallon of water, but he also has a table of jewels that are also for sale. All of the jewels on the table have a price tag of $100. You can choose between getting a gallon of water from the oasis or buying a jewel. You have to choose between either zero gallons of water or one gallon of water versus zero jewels or one jewel. The first marginal unit of water is clearly more valuable to you than the first marginal jewel. Luckily, you still have your wallet, and you have a $100 bill. You hand it over to the guard, and he gives you a gallon of water. When we use a marginalist perspective, the water is clearly more valuable than a diamond. It's only when water is plentiful that diamonds become relatively more valuable.

Marginalism extends beyond just the value of goods. Economists often refer to "thinking on the margin" in nearly every aspect of life. The logic of marginalism is intuitive to most people, but they don't necessarily realize that it is. One example of this is the idea of marginal costs and marginal benefits. Of course, everyone knows that we should do things when the benefits outweigh the costs, and we shouldn't do things when the costs outweigh the benefits. However, there's a huge spectrum. Imagine you are studying for a test. Do you study half an hour? An hour? Twenty hours? One hundred hours? A person will probably see that the expected benefits of studying for half an hour will outweigh the costs. At an hour, the benefits are still probably higher than the costs. At twenty, the additional study probably isn't doing much good, and the costs are much higher in terms of tiredness and grumpiness. At 100 hours, the benefits are probably pretty low, and the costs are very high. In short, people are "thinking on the margin" when they do some activity up to the point where the *marginal* benefit equals the *marginal* cost. Thinking on the margin is closely related to tradeoffs. Most tradeoffs involve choices about a little bit more or a little bit less, not all-or-nothing decisions.

You might have heard the old adage that "anything worth doing is worth doing well." If you agreed with that statement, then an economist might quibble with you. Instead, an economist would say that "you should do anything only to the point where the expected marginal benefits equal the expected marginal costs." The economist's saying is definitely less poetic than the old adage, but it is important to recognize that doing something well can be costly in terms of time, money, and effort. In simpler terms, thinking on the margin means that people ask themselves, "How good is good enough?" At some point, putting in more time, money, and effort isn't worth it. The question is to figure out what that point is.

Individuals think on the margin all the time, but it also applies to bigger social questions. Pollution and crime are problems, but do we really want *zero* crime and *zero* pollution in the world? An economist would say that striving for absolutely no pollution and no crime is a crazy idea. In terms of achieving zero pollution, the cleaner the air gets, the more expensive it is to clean up any remaining pollution. At some point, there will be only one molecule of pollution left in the air that we have to find. It will be extremely costly to figure out where that molecule is. The billions of dollars we would spend trying to find the last molecule of pollution would far outweigh the benefit from removing that last lonely molecule.

The same goes for getting rid of all crime. We would need constant surveillance, a huge police force, and many more courts, judges, and lawyers.

Even mundane crimes, like going one mile per hour over the speed limit, would need to be brought to justice to eliminate *all* crimes. The costs of eliminating all crimes would be enormous in terms of time, money, and effort, as well as the civil liberties we would have to give up to achieve that goal.

For all the problems that we might want to solve in the world, we need to think on the margin. Looking at the marginal benefits of some additional action versus the marginal costs tells us when we should decide that something is good enough, and then we can move on to addressing other problems. However, precisely determining the marginal costs and marginal benefits in any real-world scenario can be tricky. It is not always clear what the costs and benefits are, or we might not see the costs or benefits for a period of time. When we make decisions in groups, people have subjective evaluations of the costs and benefits, which adds another layer of complexity.

When it comes to environmental issues, policymakers, business leaders, and regular citizens constantly think on the margin. For example, a factory owner may want to reduce carbon emissions at his factory, but he must evaluate the point at which the marginal cost of reducing emissions exceeds the marginal benefit. The benefits might include the regulatory fines he would avoid or the higher profits he might make by marketing his company as "green." However, the costs might include the new technologies he must install and the retraining he would need to coordinate for his employees. The factory owner will rationally cut his factory's emissions to the point (i.e., the margin) at which the benefit equals the cost.

Another example would be how local policymakers decide to run a local recycling program. Imagine that a city already has a recycling program for plastic, paper, and aluminum, but city officials are considering adding glass recycling to their program. The city officials must weigh the marginal cost of incorporating glass recycling versus the marginal benefits. Glass is often more costly to recycle because it is heavier and bulkier than plastic, paper, or aluminum. The processing costs for recycling glass are often higher than other materials because glass needs to be cleaned, sorted by color, and crushed. Additionally, new glass is relatively cheap to produce because sand is the primary raw material.[1] Depending on the circumstances, the city officials might find that incorporating glass recycling into their program has larger costs than benefits, but it might not. Thus, the city officials don't decide between having or not having a recycling program. They decide on the margin which materials are worth including.

Now, imagine you are a billionaire philanthropist with a love for environmental preservation. You could choose to spend your money on any environmental cause, but you want to make sure that you get "the most bang for your buck"—a phrase that implies that you are thinking on the margin. You want to determine the most effective ways to help environmental preservation, so you will target your resources so that they are allocated where they generate the greatest marginal benefit. To do that, you will want to prioritize the most urgent and neglected environmental issues. For example, you might want to target the preservation of rainforests, which have some of the highest levels of carbon storage and biodiversity. Each dollar donated to the preservation of rainforests is likely to have a larger marginal benefit in terms of carbon storage and biodiversity, when compared to other types of ecosystems, like sandy deserts. Thinking on the margin helps us make incremental decisions that yield the greatest additional benefits.

4

Institutions

The word "institution" can be confusing because it has the connotation of being a physical building, or it can be very metaphorical, such as the common phrase "the institution of marriage." However, in economics, institutions are social rules that structure people's expectations and behavior, and all human action takes place within institutions.[1] Even something as simple as stopping at a red light and going at a green light is an institution that governs how we drive together on roads.

Institutions can be either formal or informal. Formal institutions are usually government-based rules, such as constitutions, legislation, and regulations. Informal institutions are rules that aren't written down but still shape what people do, such as social conventions, norms, and mores. Without institutions, either formal or informal, people would live under chaotic anarchy. Even stereotypical pirate ships from the Caribbean in the seventeenth and eighteenth centuries had very strict governance institutions to ensure that the pirates could cooperate to pillage and plunder more effectively, while also limiting in-fighting.[2]

The combination of formal and informal institutions allows people to accomplish their goals and work together effectively. For example, most people want public safety and security, but it's not always clear how to achieve those goals. We create formal rules of laws, police forces, and courts to ensure our safety and security. But the formal system can only do so much, which is why we rely on many informal institutions. Neighborhood watch programs are informal practices in which people monitor and report suspicious activities to the authorities. Mutual aid societies and charities provide assistance to people in need, which can help limit human suffering and crimes of desperation. Clubs, churches, and civic associations have regular gatherings and events that help people build a sense of community, which bolsters a collective sense of responsibility for public safety. An interconnected web of formal and informal institutions helps societies to function.

Not all institutions, however, are socially productive. Institutions are socially productive only when they channel human behavior in ways that facilitate cooperation, security, and mutually beneficial exchange. Other institutions channel behavior in socially harmful or destructive ways when they incentivize violent conflict, promote negative-sum action, and lead to instability or unpredictability.

Institutional analysis is the study of how institutions function, especially by looking at their incentive structures and their knowledge-related properties. Institutional analysis tries to figure out why some institutions can successfully fulfill their purpose, while others fail to. Institutional analysis is comparative, meaning that economists compare different arrangements to understand how they influence economic and social outcomes. Throughout much of the twentieth century, economists compare the institutions of capitalist countries to communist ones. Now that nearly all countries have given up on centralized economic planning (except for a few remaining holdouts), economists today use institutional analysis to look at different types of government policies, like the structuring of welfare, healthcare, or housing systems.

Culture is another concept that is closely related to institutions. The term "culture" has many connotations, but for many economists, culture is a socially shared and socially transmitted set of meanings. In other words, culture is a shared worldview with people around you, which often translates into similar beliefs and practices.[3] A simple example of culture is the difference between a wink and a twitch.⁴ If a space alien visited Earth and started observing us, it may see that sometimes a person will quickly close one eye. The alien might just think that closing one eye is an involuntary twitch, or it might think it is no different than a normal blink. However, people grow up knowing that a wink is not just a random twitch—it's conveying some kind of meaning. Winks might be signaling an inside joke, or some sort of playfulness, or in other cases, it might signal a flirty or romantic intention. An alien who observed a wink would have a hard time knowing what it meant, but humans know what a wink means because we have shared social meanings.

Culture can influence how well institutions actually achieve their goals. For instance, the US Constitution works relatively well within the United States because the vast majority of Americans see it as a legitimate set of rules. The average American's expectations and conceptions of good governance, individual rights, and social responsibility align with the words on the page, which has allowed those words to function reasonably well for over two hundred years. However, the US government has tried many times to force an American-style constitution on other countries, such as Iraq and Afghanistan during the Global War on Terror. Since most Iraqis and Afghans do not have the same cultural understandings of good governance, individual rights, and social responsibility, the American-style constitutions have been failures in those places. For formal rules to "stick" in the places where they are adopted, the rules need to align with the cultural understandings of the people.⁵

The world has many environmental institutions, both formal and informal. On the formal side, most countries have constitutions, legislation, and regulations that serve as the "rules of the game." A constitution is a set of meta-rules—in other words, the rules about how we make the rules—that sets up the rights that people have, how power is distributed, and how decisions are to be made. In the case of the US Constitution, it is a fairly short and vague document that outlines the basic structure of what the federal government can and cannot do. It defines the responsibilities of three branches of government and describes what kinds of legislation and regulations can be made. The Constitution also includes the Bill of Rights, which tells government officials which kinds of actions they are not allowed to do, such as limiting people's speech or engaging in cruel and unusual punishment.

Under a constitution, government officials then make the actual rules that govern a society. For example, in the United States, Congress has passed a number of environmental laws, such as the Clean Air Act, the Clean Water Act, the National Environmental Policy Act, the Endangered Species Act, the Wilderness Act, and many more. These pieces of legislation set the broad parameters, but the implementation and enforcement of the laws goes to federal bureaucracies, which includes agencies like the US Fish and Wildlife Service, the US Forest Service, the Bureau of Land Management, the National Park Service, and the Environmental Protection Agency. Usually, Congress writes relatively broad and vague laws, and bureaucrats in the agencies write specific regulations that determine how the legislation will actually operate on the ground.

For example, the US Constitution gives Congress the power to "regulate commerce with foreign nations, among states, and with the Indian tribes." This rule has been interpreted loosely, and Congress used its power under the Commerce Clause to pass the Endangered Species Act.[6] This law does not specify which species are to be listed, but it gives the US Fish and Wildlife Service and the National Marine Fisheries Service guidelines for how to list a species as endangered, as well as guidelines for how to protect them. These agencies then make specific rules for the day-to-day implementation of public policies that are intended to save endangered species. In all, the preservation of endangered species involves formal institutional rules at a variety of levels, including constitutional, legislative, and regulatory.

In addition to formal institutions, informal ones play an important role in addressing environmental issues. One example of this is the social shift regarding the wearing of animal furs, such as sable, mink, chinchilla, and lynx.

Over the course of the twentieth century, wearing animal fur garments was an important and conspicuous status symbol. However, in the mid-twentieth century, the anti-fur consumption movement heavily stigmatized the wearing of fur. Animal rights organizations like People for the Ethical Treatment of Animals helped to shift social norms, as informal institutions, so that furs were no longer seen as luxurious or glamorous, but instead, the wearing of fur became scornful and shameful. Additionally, technological advancements made faux fur look authentic, providing a viable alternative that didn't directly harm living animals.[7] Social stigma is often one of the most powerful informal institutions for shaping human behavior.

5

Property Rights

One of the most important institutions is property rights because they define the privileges that individuals have with specific resources. Property rights can be protected formally by law, or informally by social norms. In other words, property rights make clear who can make decisions about a particular resource, and they assign liabilities or costs for harming other people's resources. If you harm other people's property rights, you might be fined, receive jail time, face social ostracism, or lose your reputation. Property rights align the incentives of owners to make prudent decisions about what they own because they reap the benefits and bear the costs of their choices.

Property rights aren't just one thing. They consist of a "bundle" of many sub-rights that can be separated. Some of these sub-rights include the ability to own something, sell it, lease it, exclude others from it, use it, leave it unused, generate profits from it, protect it from others, and seek redress from harms against it. For example, think of buying your own home. Homeowners have broad freedom—but not unlimited—to use their property as they see fit. They could plant an orchard near their house. They can use their property as collateral for a loan. They could rent their house to someone else. They can keep other people off their property, with the help of the police. If a neighbor dumps trash in their backyard, they can bring the neighbor to court for violating their rights. If the government tries to take their land through eminent domain, they can challenge the taking or gain fair compensation for their loss.

However, people can willingly and contractually choose to give up some of their rights. Many people live in homeowners' associations (HOAs), which are restrictions on the rights of homeowners. If you choose to buy or build a house in an HOA, you may not have the freedom to paint your house whatever color you would like, and you may not have the freedom to let your grass grow as long as you want. HOAs are intended to keep the property values high and maintain a certain "vibe" in a neighborhood, but the tradeoff is that homeowners choose to give up certain sub-rights. Some HOAs are stricter than others, and many neighborhoods don't have HOAs at all, so people can choose which types of property rights most closely fit their preferences.

Property rights come in four general categories. First is individual private property, in which a single person gets to make decisions about how the property is used. Second is communal property, in which a group of people jointly own something, and they have to make collective decisions on how to use it. Third is public, or government, property, in which political actors make decisions for property that is then administered by government bureaucrats. Fourth is an open-access resource, which is not owned by anyone.

There is not one type of property system that is inherently better than another. For instance, if you care about American bison, you can see them in two types of places: public property and private property. Large bison herds are in Yellowstone National Park, which is public property. Congress makes laws for how national parks and their wildlife should be managed, and the National Park Service implements those laws and makes specific policies and regulations based on the guidance in the laws.[1] Millions of people go to Yellowstone each year to see the bison herds. But, bison herds are also privately owned on private land. Ted Turner, the billionaire founder of CNN, owns several large ranches in the western United States, including Flying D Ranch in Montana. Flying D Ranch is over 113,000 acres, and one of Turner's goals is to produce bison while also serving as a wildlife refuge. Turner also sells trophy bull bison hunts on his land. Both the public property of Yellowstone National

Park and the private property of Ted Turner's Flying D Ranch have facilitated larger bison numbers, just by using different means.[2]

There are a wide variety of institutional means to protect property rights. In many cases, a government uses its coercive power to define and enforce property rights with police and courts. Although it is common for governments to be the main protector of property rights today, that was not the case historically when governments were much less powerful and far-reaching. Even today, societies in the hinterlands have many effective nonstate forms of property rights. In places today where government is effective, officials cannot be everywhere at once, so other forms of nonstate property rights protection exist. Some common examples of nonstate protection are surveillance systems and alarm services that deter theft or vandalism. Community norms of monitoring and enforcement, such as neighborhood watch programs, limit the number of burglaries. Cultural norms and expectations of respecting property usually result in social sanctions and ostracism for violators. Thus, governments, individuals, and communities have many means to protect property rights.

Property rights, particularly private ones, are socially important because they incentivize people to create, innovate, and conserve their property. Private property rights are also socially beneficial because they allow voluntary trading among willing parties, and trading among people allows resources to flow to their more highly valued uses. Property rights tie the rewards and penalties of entrepreneurial behavior. In other words, owners tend to maximize the value of their assets.

One of the clearest examples of this is the difference between animals that are owned (like livestock) versus animals that are unowned (many kinds of wildlife). Every day, millions of livestock, such as cows, pigs, sheep, goats, chickens, and ducks, are slaughtered for food or used to produce other goods, like milk or wool. Despite the fact that millions are killed each day, these species are not in danger of extinction. The farmers and ranchers who raise livestock have a strong incentive to ensure that the source of their livelihood is sustained. They also have a strong incentive to protect their livestock from anyone who would steal or kill it. However, in many places around the world, it is illegal to own wildlife, and many of these species currently face extinction. When wildlife is either unowned, or owned by a government, regular people have weak incentives to propagate the species or to go after poachers.

Most environmental problems arise when property rights are not clearly defined, not well enforced, or not easily exchanged. One example of this problem is the "tragedy of the commons." In the traditional sense, a commons is a shared piece of land where a group of farmers have the right to graze their animals without fences or boundaries. None of the farmers can stop the others from grazing their animals, and no particular farmer owns the land. They own it in common. Thinking logically about the incentives of each farmer, they will want to put their animals on the commons to graze as quickly as possible because if they wait, there might not be any grass left. Since each farmer understands this logic, they will all put their animals on the commons as quickly as possible, and the resource—in this case, the grass—will be depleted.

The tragedy of the situation is that the rationally self-interested actions of each of the farmers lead to a socially undesirable outcome that harms everyone in the group. The logic of the tragedy of the commons can extend to any resources that are owned by no one, or by society at large. These resources tend to be overused, misused, and abused because an individual has weak incentives to work towards the collective good.

Although the tragedy of the commons can happen, in the real world we see that many groups of people create rules to govern these "common-pool resources" so that people can cooperate and avoid the tragedy. For example, overfishing is a big problem in the world's oceans because the open oceans are unowned and so are the fish. This is a problem because the benefits go to each person, but the costs are shared by everyone, meaning that individuals have a strong incentive to take more than they should.

To avoid the "tragedy" of overfishing, governments in some places have put in market-like regulations called individual fishing quotas (IFQs), also known as catch shares. In this system, the government first caps the total sustainable amount of fish that can be caught in a particular place. Then it divides up that total sustainable amount into shares, and it hands out those shares to specific fishermen. The IFQ tells a person how much they are allowed to catch. Each fisherman's catch share becomes like a form of private property, which can be traded among different fishermen. Thus, the IFQ system is a cap-and-trade system that has effectively ended the tragedy of the commons off the coasts of the United States and Canada, among many other places.[3]

However, since no government has jurisdiction over the open ocean, the tragedy of the common still occurs, but there are alternative approaches that could help address it. One potential solution is aquaculture, which is the raising of fish in "fish farms" in the ocean. Large nets create "tanks" in the ocean, so fish can be raised like other livestock. Thus, aquaculture can increase fish production while reducing pressure on wild stocks. Norwegians are the leading international producers of Atlantic salmon through aquaculture.[4] Of course, no solution is perfect, and there are always tradeoffs. In these fish farms, pollution and disease can be a real issue. However, although the existence of pollution and disease from fish farms isn't ideal, such an outcome may be a better outcome than the collapse of open-ocean fisheries.

A tragedy of the commons doesn't always happen on large scales, like open oceans. In fact, many problems with such resources are fairly small in scale. When people collectively own a common-pool resource, it is difficult to exclude people from using it because everyone has a claim. However, the resource can be easily depleted by overuse. To avoid the tragedy of the commons, people must figure out rules for sustainable use. Evidence shows that many local communities have developed creative rules to manage their common-pool resources without falling into the tragedy. Elinor Ostrom—the first woman to win the Nobel Prize in economics—became famous for her work on how people in the real world ingeniously overcame the tragedy of the commons.[5] In her research, she found that local communities devised rules to ensure the sustainable extraction of commonly owned resources. Just a few examples that she studied were the management of alpine meadows in Switzerland, irrigation systems in Japan, fishing areas in Turkey, and forests in Nepal. Full government control and privatization weren't necessary to avoid the tragedy, as many scholars before Ostrom had presumed.

Another important concept related to property rights is externalities, also called spillover effects. In the simplest terms, an externality is a situation in which one person's actions affect another unwilling person in some way. Externalities can be both positive and negative. For example, a positive externality is the delicious aroma you smell as you walk past a bakery. You didn't pay the bakery for the enjoyment you experienced, and you didn't ask for it. A common negative externality is air pollution. If your neighbor decides to burn some tires in his backyard and you are downwind, you will have to deal with the black smoke and the stench of burning rubber. Your neighbor's unwanted actions are spilling over onto you.

However, the previous explanation of externalities is too simplistic. Most negative externalities are not always so straightforward, and they only become clear when property rights are defined. Imagine you live in an apartment beneath a neighbor who is just learning the drums. Your neighbor is at work all day, so the only time he has to practice is at night when you are trying to sleep. You get frustrated with all the noise, so you loudly jab a broomstick at the ceiling, put passive-aggressive notes in his mailbox, and tell him to be quiet when you pass him in the hallway. Is your neighbor's drumming a negative externality *to you*, or is your incessant complaining a negative externality *to him*? When property rights are not clearly defined, the externality occurs in both directions. In other words, without some conception of property rights, the externality is reciprocal. If the apartment management clearly specifies quiet hours, then you have a right to a peaceful night's sleep, and your neighbor is imposing a negative externality on you. If the apartment is managed by music enthusiasts who encourage practicing, then your drumming neighbor has the right to play, and your complaining is a negative externality on him.

Despite the ambiguity of externalities in many scenarios, people can find ways to bargain among one another to overcome perceived externalities, even if property rights are not established. The ability to bargain over externalities is called the "Coase Theorem," named after economist Ronald Coase. The Coase Theorem says that when two parties have a conflict over property rights or externalities, they can bargain to an economically efficient outcome, no matter who had the initial property rights, as long as transaction costs are sufficiently low.[6] In the simplest terms, the Coase Theorem says that people can solve conflicts over property rights and externalities by bargaining with one another, but let's unpack the ideas a bit more.

A transaction cost is any cost that is involved in engaging in an exchange, which could take several forms, like money, time, or inconvenience. Transaction costs include several sub-categories of costs. Search and information costs are those that people incur when they try to identify possibilities for mutual gains or the costs of gathering information to make informed choices. Bargaining and decision costs are those that people incur when they try to negotiate an agreement, like the time spent in meetings and the time used in writing contracts. Policing and enforcement costs are those that people incur when making sure that everyone sticks to an agreement. In the context of the Coase Theorem, "economic efficiency" means that the allocation of resources maximizes total social welfare or value, accounting for all the costs and benefits of everyone involved. Maximizing social welfare means that the total benefits to everyone are as high as possible, accounting for all the various costs that people incur.

In the end, when transaction costs are low enough, people can bargain so that property rights flow to the people who value them the most, thus maximizing social welfare. An example can illustrate the Coase Theorem in action. Imagine a small town that contains a few hundred houses and a giant factory. The factory belches out black clouds of pollution every day, which harms the health of the residents and makes their town uglier. Let's assume the town has no laws against air pollution, and let's assume that the residents don't have a right to clean air, and the factory doesn't have a right to pollute. What can be done? One option is for the factory to act as if the residents have the right to clean air. The factory can then bargain or negotiate with the residents for the right to pollute. The factory might compensate each of the residents for the harms that they suffer, like their medical bills. It will cost the factory $30 million to compensate the residents to fix their harm. The second option is for the residents to act as if the factory has the right to pollute. The residents can then negotiate with the factory to reduce pollution. They can pay the factory to implement cleaner technologies or reduce production. It will cost the residents $10 million to compensate the factory to the point where the pollution is acceptable to them. If the transaction costs are low enough, the negotiations between the factory and the residents lead to an outcome where the total welfare of everyone involved is maximized. In this simplified scenario, the negotiations will lead the residents to pay the factory $10 million because it is the lowest-cost solution to the problem.

The simple scenario above may not seem very realistic, and it isn't in a lot of ways. In the real world, transaction costs very much exist. The implication of relatively high transaction costs is that the initial assignment of property is incredibly important because bargaining can't easily happen.[7] Despite the problems of transaction costs, people in the real world find creative and innovative ways to bargain with one another all the time. The world may not fit the Coase Theorem exactly, but it can get fairly close to it in many cases.

6

The Entrepreneurial Market Process

When economists use the term "the market," sometimes they are referring to a physical space where people buy and sell goods, like a weekend farmer's market. Sometimes the term is an abstract metaphor for all buying and selling of a particular good. Additionally, economists refer to the market as a process, which might be the most unfamiliar use of the term. The "market as a process" concept is meant to capture all the constant adjustments of people's actions and interactions through time and space. The world is a constantly changing place. Some resources become more plentiful, and others become scarcer. People's preferences change over time. New discoveries are made. Disasters strike. The market has a series of feedback mechanisms that help both buyers and sellers to learn and adjust, based on these changing conditions—making the market a process.

Many of the insights of the market as a process go back to Adam Smith in *The Wealth of Nations*. Smith realized that economic growth arises from people dividing their labor, specializing, and engaging in mutually beneficial exchange. The division of labor and specialization are two of the most important insights from Adam Smith. Imagine you were shipwrecked on a deserted tropical island, and you have to survive on your own. You'll have to collect fresh water, build shelter, start a fire, and find food, all of which can be time-consuming and difficult. If you were truly alone, you could spend all day working, and you would still have very little fresh water, a relatively ramshackle shelter, almost no food, and maybe a modest fire, if you're lucky enough to know how to start a fire without matches.

However, now imagine that you were shipwrecked with thirty other people. You could organize yourselves so that you each took on different tasks of surviving. By dividing up your labor and specializing in particular tasks, each of you can become proficient, and maybe even excellent, in those tasks. The people who collect fresh water could collect much more water if that's all they have to focus on. The people who build shelters would learn how to build sturdier shelters for everyone to live in. The people who gather food could specialize even further, with some people perfecting their ability to fish, while others improve on gathering coconuts. Everyone can then trade with one another, and everyone is wealthier than they would otherwise be if they had to do everything on their own.

Adam Smith said the division of labor is limited by the extent of the market, which means that as we are in bigger societies, the larger number of people specialize more and engage in more trades, which makes us even wealthier. In today's globalized economy, the extent of the market is roughly eight billion people, making us richer and more technologically advanced than ever before. When populations are isolated, they have fewer people, which means fewer minds and fewer hands to find new and better ways to produce things.

Markets have five major components: private property, knowledge, prices, entrepreneurs, and profit-and-loss. The combination of these five components allows the market process to function. Since conditions are always changing, people need to readily adapt. In other words, consumers want a variety of goods and services, but they want them in the correct quantities and qualities, and at the correct times and in the correct places. However, the "correct" quantities, qualities, times, and places can and do change every moment. Consumers will have very different thoughts about what they want the day before a hurricane hits and the day after. The big question is how the whole market system adapts and responds to all the constant change going on around us.

The foundation of the market process is private property, particularly property that is secure and can be easily traded. Property owners can gain by using their resources in ways that are beneficial to others. In economics jargon, property owners are "residual claimants," meaning that they get to claim the residues—profits or assets. Since people reap what they sow, they have an incentive to care for and manage what they own for their own benefit. On the other side, property rights assign liabilities, so people have an incentive to limit the damage or harm to other people's property. Private property provides the basis of the division of labor and specialization that brings about economic growth and human flourishing.

Knowledge is the next component of the market process, and it comes in many forms. First is scientific knowledge, which is usually relatively objective, and often possessed by experts. However, some of the most important knowledge is both dispersed and local, subjective and context-dependent, situated in a specific time and place.[1] Dispersed, local knowledge is a practical kind of knowledge that people get through their own experiences. For example, you know the local conditions of your neighborhood and your workplace. You have knowledge about who to trust and who to avoid. You have knowledge about the social networks that you interact with. This kind of everyday, on-the-ground knowledge is dispersed among millions of individuals, and there is no effective way that such knowledge can be fully aggregated or centralized. In many ways, this type of on-the-ground, practical knowledge can't be turned into digitizable data for computation.

Relatedly, many kinds of knowledge are tacit and inarticulable, meaning that you know you know something, but you can't really say what you know.[2]

For example, think of riding a bike. You can try to teach someone else how to ride a bike by explaining it to them. You could say that you sit on the seat, hold the handlebars, and push the pedals with your feet. However, if you've seen anyone, a child or an adult, try to learn how to ride a bike, telling them how to do it doesn't do much good. Unless a first-time bike rider has training wheels, they are likely to tip over quickly, even if you explained how to ride the bike as clearly and with as much detail as possible. That's because riding a bike successfully is based mostly on an intuitive sense and a felt understanding that is built up with experience, rather than something that can be explained verbally.

One of the problems with knowledge is that we don't know what we don't know. This is called "sheer" or "radical" ignorance. In other words, there is a difference between "known unknowns" and "unknown unknowns." I know that I don't know how to fly a commercial aircraft, but I realize that if I went to flight school, I could learn. Radical ignorance is a much more difficult problem because we can't comprehend the scale or scope of things that we don't know. All people are constantly unaware of opportunities, possibilities, or even problems that can and do exist.

Relatedly, there is an important distinction between "risk" and "uncertainty."[3] As economists talk about it, risk is a calculatable thing. If I roll a pair of dice, I know that I have a 1-in-36 chance of getting snake eyes. So, if I'm gambling, I know the risk of my bet. However, uncertainty refers to a situation when the potential outcomes are unknown, and no probabilities can be assigned. Uncertainty means we're truly in the dark.

In a world of uncertainty and radical ignorance, we need mechanisms of discovery and adjustment so that we can update our actions as we learn new things. The mechanisms of the market process facilitate discovery and adjustment, as will be explained below.

The next component of the market process is prices. When people have private property, they begin to trade with one another. When large numbers of people begin to trade, market prices emerge, which are simply exchange ratios.

In other words, money prices serve as a common denominator so that we can compare between things that are seemingly unlike. It's important to emphasize that prices aren't just arbitrary numbers. Instead, they communicate at least three important things: dispersed knowledge, the relative scarcity of goods, and opportunity cost. When a good's price rises, that is a signal that consumers should economize on their use of that good. It also acts as a signal that producers should produce more or find alternatives to provide consumers with.

Hayek described the role of prices in terms of aggregating dispersed knowledge and inducing people to change their behavior based on changing conditions. He said to imagine a world in which some good is scarcer than it used to be. We can use the example of gasoline. Maybe there is some new, valuable use for gasoline that didn't exist before, so people are racing to use more. Or, maybe a major gasoline production site was shut down, meaning that there is less gasoline around to use. Hayek emphasized that market prices are important communicators of knowledge because consumers of gasoline don't need to know the specifics of why it is scarcer. They don't need to know whether a new use for gasoline has been developed, or whether a major oil refinery has shut down. As Hayek argued, all that the users of a good like gasoline need to know is that what they used to consume is now more profitably employed elsewhere and that they must economize on their use of that good. Consumers of gasoline don't need to know the specifics about its production. All they need to know is that the price has gone up, which tells them that they need to change their behavior. Relatedly, producers don't need to know all the specifics about changes in the gasoline supply. The higher price tells producers that they should either produce more or find viable substitutes. The effects of a price change rapidly spread throughout the whole economic system, which influences not only all the uses of the good, like gasoline, but also those of its substitutes.[4] The adjustments happen because prices communicate the most important things that producers and consumers need to know—whether to produce or consume more or less.

With private property and market prices, the next element of the market process is entrepreneurship. An entrepreneur isn't necessarily a special kind of person; instead, anyone can act entrepreneurially. In economics, two general conceptions of entrepreneurship exist: alertness and innovation. Alertness is being attuned to previously unknown and unexploited opportunities for profit. Since the world is always changing and new opportunities are constantly emerging, entrepreneurs are the people who recognize new ways of doing things. The simplest version of this is buying something at a low price in one place and selling it for a higher price in another place. However, entrepreneurs can be alert to and discover many types of things: new goods, new production methods, new potential markets, previously unused resources, new ways of organizing, and so on.[5]

Once entrepreneurs have been alert to and discovered something, they must seize that opportunity. They put their ideas into action, which makes them agents of innovation, or as the economist Joseph Schumpeter called it, "creative destruction." Although the term may sound drastic, creative destruction refers to the process of replacing older, inefficient goods or means of production with better, more efficient ones.[6]

Entrepreneurs cater to the values and desires of potential customers. Since entrepreneurs are rationally self-interested, they see that they can benefit by making consumers happy. If consumers care deeply about the environment, then entrepreneurs have an incentive to discover new opportunities for being ecofriendly and selling those products to consumers. Of course, this depends on the degree to which consumers care about ecofriendliness, but if enough consumers with such preferences exist, an entrepreneur can profit by strategically catering to those preferences. There are many examples of entrepreneurs who cater to the environmental preferences of consumers, leading to the creative destruction of more ecologically harmful goods to less harmful ones. The electric car industry is booming, replacing many fossil-fuel-powered cars. Many consumers demand clothing that is produced with sustainable materials, and some clothing manufacturers have responded by using recycled materials in their production processes. Plant-based meat substitutes and lab-grown meat are two ways in which entrepreneurs are providing alternatives to conventionally produced meat products. Some entrepreneurs are growing fruit and vegetables organically to avoid the environmental effects of conventional agriculture, catering to the preferences of eco-conscious consumers.

After entrepreneurs have discovered a new idea that they think consumers want, they need some sort of feedback to know the degree to which their idea is worth pursuing. Within markets, the system of feedback and discipline is profit and loss. Entrepreneurs expect that their discoveries of previously unrecognized profit opportunities will be successful, but there is no guarantee. Entrepreneurs might have been correct in their expectations, or they might have been totally wrong. Profits are a general signal that entrepreneurs' choices are fulfilling consumers' desires in a more effective or efficient way than the alternatives. Losses are a general signal that entrepreneurs' choices are not fulfilling consumers' desires as well as some other course of action. Entrepreneurs are lured by the potential for profits, which prods them to seek innovations that give consumers the things they want, and they are disciplined by the threat of losses, which induces them to behave prudently. However, profits and losses are only useful feedback mechanisms if markets are relatively free. When government policies distort the way that markets work, some actions that would be profitable are not, and actions that wouldn't be profitable become so.

Entrepreneurship certainly exists in the market, but it also exists in the political and social spheres. Political entrepreneurs look for and exploit potential political profit opportunities, which might include things like different policy approaches or facilitating connections between influential people.[7] Social entrepreneurs use social capital and networks to facilitate social change, often creating civic associations or nonprofit organizations to accomplish their goals.[8] Political or social entrepreneurs can help to solve many social problems, but there is also the potential that they exacerbate them. For example, some political entrepreneurs could champion a new policy that subsidizes the research and development of carbonless energy sources, making it easier to combat climate change, but other political entrepreneurs could cater to influential companies and implement policies that subsidize fossil fuel extraction. Likewise, a social entrepreneur could create a nonprofit organization to advocate for the expansion of renewable energy production on public land, but an opposing social entrepreneur could create an organization that seeks to bar any such development.

Unlike market entrepreneurs, the feedback mechanisms for political or social entrepreneurs are a bit looser and less clear cut. Political and social entrepreneurs lack prices and profit-and-loss in their decision-making processes, so their feedback mechanisms are things like election results, community support, etc.[9] With these looser feedback mechanisms, the social helpfulness or harmfulness of political and social entrepreneurs depends on the broader institutional rules in society, which will either shape the behavior of these entrepreneurs in productive or destructive ways.

The market process seems complicated, so a person might ask themselves, "Why not just do central planning instead? If we just had intelligent people to run the economy, couldn't we skip the messiness of the market process and centrally allocate all the goods that people need?" Since the rise of the Industrial Revolution, some people have been outraged by the inequality that arises in market systems, so they wanted to use rational centralized planning as a more desirable economic system without those flaws. One of the first large-scale attempts at central planning was in the Soviet Union. Other places followed the Soviet lead, such as Maoist China, Vietnam, Cuba, North Korea, Poland, Hungary, Czechoslovakia, and other Eastern European countries. However, central economic planning failed, and it failed catastrophically, but why?

The problem is that relevant knowledge couldn't be discovered, aggregated, communicated, and adapted to in the same way that the entrepreneurial market process allows. Additionally, central economic planners and the workers in the system had much weaker incentives to produce the goods and services that people demanded. To make matters even worse, the main motivations for central planning were to overcome inequality and instability, but these centrally planned economies couldn't solve those issues either.[10]

The market process needs to be messy, and it can't be centrally planned because rivalry is what allows us to engage in knowledge discovery and adjust to changing conditions. As economist Don Lavoie wrote,

> Economic rivalry is the clash of human purposes. It is the aspect of market relations that is revealed, for example, every time one market participant bids away resources from another. When one competitor undercuts the price of arrival; when one consumer buys the last retail item in stock before another consumer gets there; when one inventor beats another to the punch on a profitable innovation—that is economic rivalry.[11]

The push and pull of all the people in the market allows each person to adjust their particular plans with the plans of billions of other people, as well as the unceasing dynamism of the real world. As Lavoie continues,

> The entrepreneurial market process requires certain forms of rivalrous activity, such as outbidding one's competitor, but yields extremely beneficial results: It generates the continuously changing structure of knowledge about the more effective ways of combining the factors of production. This knowledge is created in decentralized form and dispersed through the price system to coordinate the market's diverse and independent decision makers.[12]

In sum, one of the most important aspects of markets is that they are rooted in positive-sum arrangements because participants voluntarily choose to engage in them. A positive-sum arrangement necessarily means that both parties voluntarily agreed to terms that make them better off, at least in their own subjective valuations. However, being a positive-sum arrangement doesn't mean that both sides benefit to the same degree. Even if a voluntary exchange benefits one side more than the other, the positive-sum outcome is still better than a zero- or negative-sum outcome. When situations are positive-sum, people are more likely to engage in cooperation and peaceful negotiations, meaning that markets are composed of "millions of mutually beneficial nonviolent actions."[13]

7

Supply and Demand

One of the most basic and helpful tools in economics is the idea of supply and demand. Supply is the combined amount of a good that all producers in a market are willing to sell. Demand is the combined amount of a good that all consumers are willing to buy. Economists show this concept graphically, which is a simplification and abstraction of the complex processes that actually happen in a market.

First, there is the downward sloping demand curve, which shows the relationship between the quantity demanded by consumers and price, holding all other factors constant. From the curve, you can see that as the price of a good goes down, consumers will buy more of a good. In other words, a higher price means fewer goods are bought, and a lower price means that more goods are bought. In Figure 7.1, you can see the full demand curve for a hypothetical market. In this example, at a price of $300, the quantity demanded will be 2000 units of the hypothetical good.

Second, there is an upward sloping supply curve that shows the relationship between the price and the quantity supplied by producers. From the curve, you can see that as the price of a good rises, producers will supply more of that good. In other words, a higher price means that more goods are produced, and a lower price means that fewer goods are produced. In Figure 7.2, you can see the full supply curve for the same hypothetical market. At a price of $300, the quantity supplied will be three thousand units.

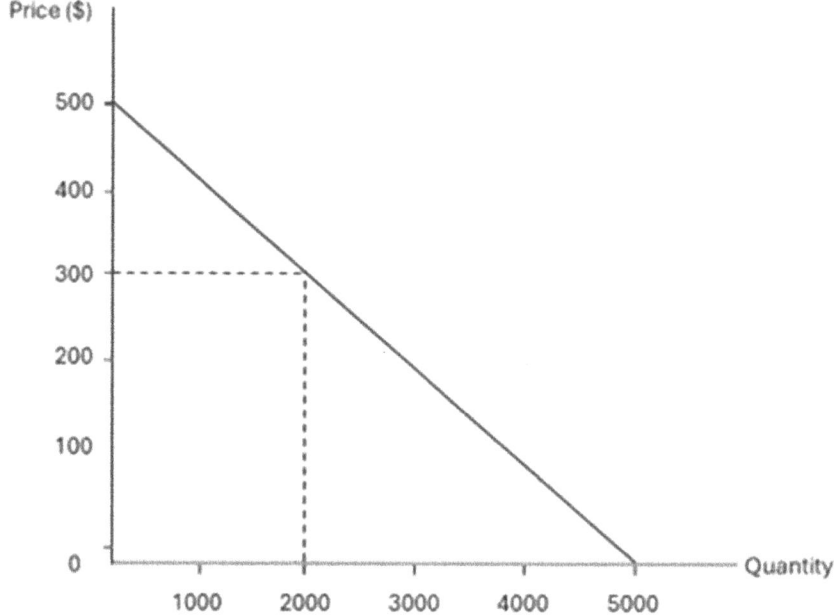

FIGURE 7.1 *Demand curve*

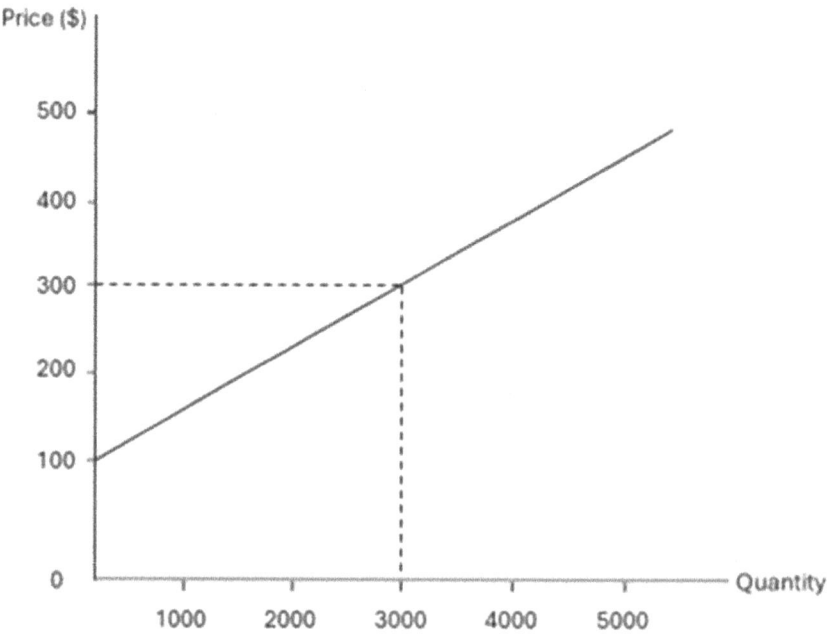

FIGURE 7.2 *Supply curve*

When we put the two curves together, we can see a market equilibrium in the hypothetical market. The place where the two curves intersect shows the equilibrium price and quantity of goods (see Figure 7.3). But you may be wondering why the equilibrium exists. The interactions between the buyers and sellers determine the equilibrium price, which is where the quantity demanded is equal to the quantity supplied. At any other price, the actions of entrepreneurs and consumers push the market toward the equilibrium price and quantity. In a market, buyers compete against other buyers, and sellers compete against other sellers. A buyer gets the good she wants by bidding higher than other buyers. Sellers try to offer their goods at lower prices, trying to undercut one another to woo buyers. For example, at an auction, the buyer with the highest bid gets the item, and the seller with the lowest price gets the sale. In Figure 7.3, the market equilibrium is shown, with a price of $250 and a quantity of 2,200 units.

Imagine that the price is set higher than the equilibrium price. The quantity demanded is lower than the equilibrium, and the quantity supplied is higher.

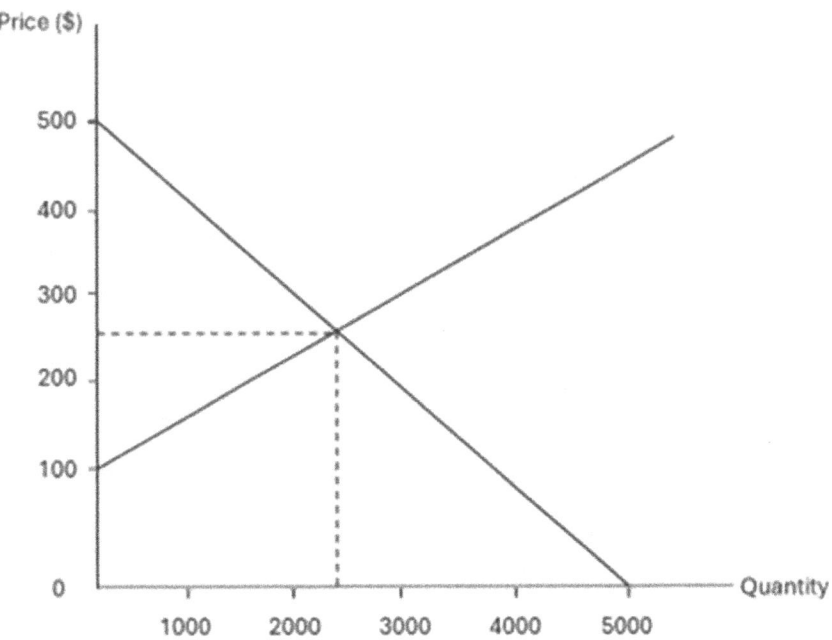

FIGURE 7.3 *Market equilibrium*

In this case, there will be too much of the good on the market, which we call a surplus. Each seller sees an entrepreneurial opportunity to lower their price, if possible. Each seller tries to undercut the other sellers by lowering the price of the good. As sellers start to drop their prices, buyers start to buy up more of the good. The point at which sellers stop dropping their prices is at the point where the quantity demanded is equal to the quantity supplied.

On the flip side, if the price is lower than the equilibrium price, then the quantity demanded is higher than the equilibrium, and the quantity supplied is lower, meaning that there is a shortage of the good. When there's a shortage, buyers can't get as much of the good as they would like at the current price, and sellers aren't willing to sell at such a low price, likely because they can't afford to. Buyers bid up the price, and sellers have an incentive to raise the price since buyers are willing to pay the higher price. The market price rises until quantity demanded equals the quantity supplied. At any price other than the equilibrium price, the incentive of buyers and sellers is to push the price toward the equilibrium price.

At equilibrium, not all buyers will buy, and not all sellers will sell. The buyers to the right of the equilibrium point won't buy because the price is higher than they value the good, but the buyers to the left of the equilibrium value the good more than the price. Similarly, all the suppliers to the right of the equilibrium point are nonsellers because their costs of production are higher than the equilibrium price. All the suppliers to the left are sellers because their costs of production are lower. Since the buyers who place the highest value buy first, and the sellers with the lowest costs sell, then there are massive gains from trade. In other words, the difference between the value that a good creates for a consumer and its cost to produce is maximized. Every trade that generates value will generate value until the last trade where the last buyer gets the smallest amount of value higher than the cost. The equilibrium point is exactly where the value to buyers is equal to the cost to sellers (see Figure 7.3).

In this idealized, simplified theory of markets, buyers and sellers fully coordinate with one another so that there are no unexploited gains from trade, and there are no wasteful trades. If the quantity exchanged between buyers and sellers were greater than the equilibrium quantity, then very costly goods would be produced for relatively frivolous desires. Markets are based on buyers and sellers acting purposefully in their own self-interest, which leads to a price and quantity that allocates goods to the buyers who value the goods most highly, while also being produced by the sellers who can do so at the lowest cost. This combination maximizes the gains from trade, meaning that the total of the benefits to both buyers and sellers is as high as it can be. Of course, real-world markets are much messier and more complicated than the idealized theory suggests, but we observe that markets seem to tend to an equilibrium. To make matters even more complicated, the real world is highly dynamic, meaning that a market equilibrium is always changing. Again, an important caveat is that supply-and-demand analysis is helpful for understanding the world, but it doesn't perfectly describe the nuances of the real world.

The logic of supply and demand works in a simple, idealized scenario, but it is also helpful for thinking about the effects of government policies that artificially change the price or quantity. For example, we can think of a price floor, or a mandated minimum price, like a minimum wage law. If a law forces the price higher than the market's equilibrium price, then there will be a surplus because there is a gap between the quantity that sellers are willing to supply and quantity that buyers are willing to pay for. In the case of a minimum wage, the demander is an employer, who wants to buy labor. The supplier is the worker who is selling their labor on the market. Since the price is artificially high, there are more sellers (workers) than buyers (employers), and the surplus is a surplus of workers—meaning that there is some degree of involuntary unemployment (see Figure 7.4).

The reverse logic is also true. An example of a price ceiling, or a mandated maximum price, is a rent control. Under these laws, landlords are not allowed

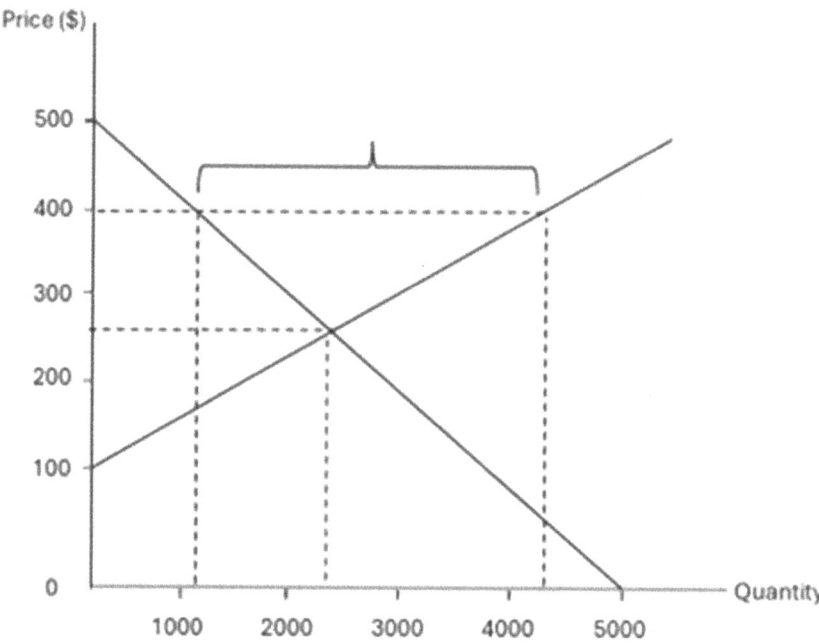

FIGURE 7.4 *Surplus caused by an artificially high price*

to charge a price that is higher than the market would otherwise permit. When a law forces the prices lower than the equilibrium price, there will be a shortage. There will be more demanders (people who want to rent an apartment) than sellers who are willing to sell (landlords who are willing to rent their apartments) (see Figure 7.5).

In the real world, the underlying determinants of supply and demand can change, meaning that the supply curve or the demand curve for any good can shift. For the supply curve, there are a variety of factors that might cause it to shift: the cost of production can change due to technological innovations; the number of sellers can change; the sellers' outside options can change; or there might be some sort of external shock like a natural disaster. For the demand curve, changes in the various factors might cause it to shift, like the number of consumers, consumers' preferences, the prices of other goods, or consumers' wealth level. For example, if an unexpected cold snap suddenly wiped out 50

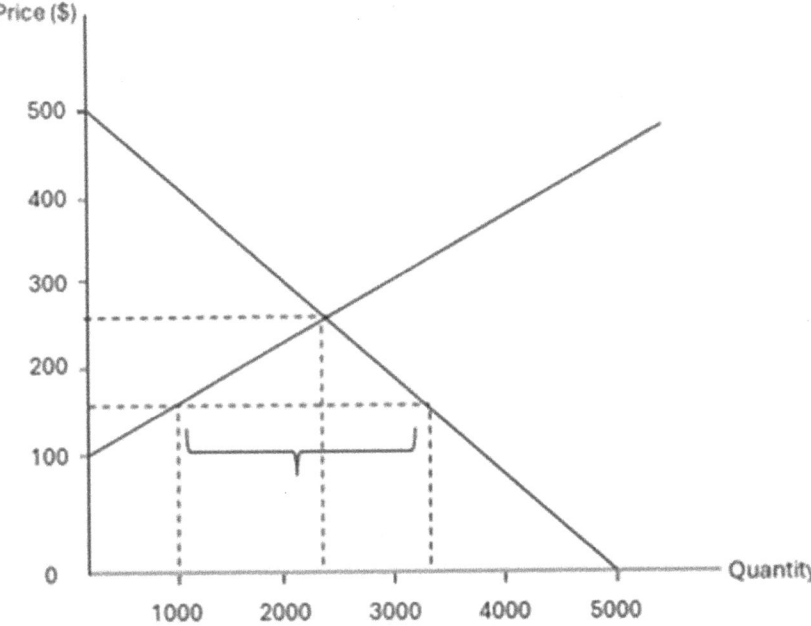

FIGURE 7.5 *Shortage caused by an artificially low price*

percent of all the strawberry crop in a country, then the supply curve would shift to the left, leading to a higher equilibrium price for strawberries. The logic is the same on the demand side. If 50 percent of people decided to become vegans, then the demand curve for beef would shift to the left, meaning that the equilibrium price for beef would decrease (see Figures 7.6 and 7.7).

The logic of supply and demand is a simplified view of the world that is not fully realistic. However, the logic is still useful. We can make general predictions about how humans will behave in different scenarios. For example, the purpose of a government-granted subsidy is to make a good cheaper than it would otherwise be on the open market. We know from the logic of supply and demand that the subsidy will artificially lower the price, meaning that the quantity demanded will increase. This may be a problem, for example, if a government subsidizes water usage in an arid area, leading consumers to use more water than they otherwise would have.

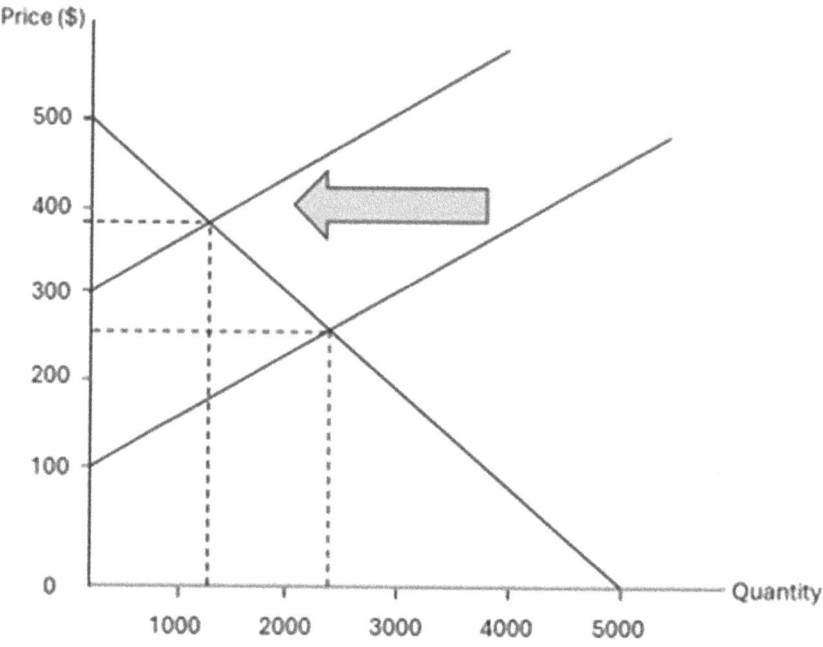

FIGURE 7.6 *Shifting the supply curve due to a decrease in supply*

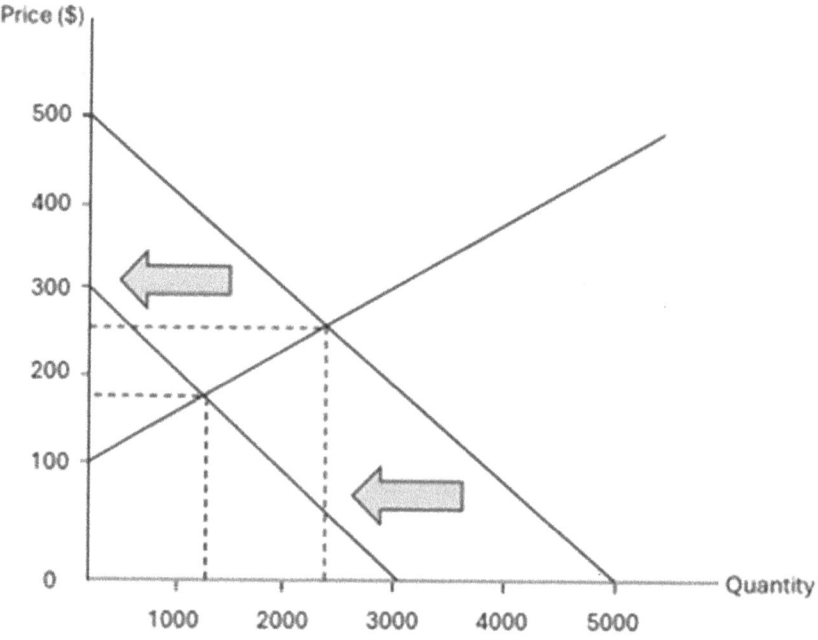

FIGURE 7.7 *Shifting the demand curve due to a decrease in demand*

8

Unintended Consequences

It's no surprise that actions have consequences, and some of those consequences are unintended. One of the first economists to talk about unintended consequences was Frédéric Bastiat, a mid-nineteenth-century French scholar, politician, and satirist. He framed the concept as "the seen" and "the unseen." As Bastiat wrote,

> In the sphere of economics an action, a habit, an institution or a law engenders not just one effect but a series of effects. Of these effects only the first is immediate; it is revealed simultaneously with its cause, it is seen. The others merely occur successively, they are not seen; we are lucky if we foresee them. The entire difference between a bad and a good Economist is apparent here. A bad one relies on the visible effect while the good one takes account both of the effect one can see and of those one must foresee.[1]

What Bastiat is saying is that actions have easily visible effects, which are relatively straightforward. However, the most important thing is to try to foresee the potential aftereffects that will arise from an action. Another one of Bastiat's stories, known as the "Broken Window Fallacy," shows this example. He imagines a scenario where some townsfolk see a boy break a window. It may seem strange, but they are happy because "Good comes out of everything. Accidents like this keep production moving. Everyone has to live. What would happen to glaziers if no window panes were ever broken?" Then Bastiat describes that the "glazier will come, do his job, be paid six francs, rub his hands and in his heart bless the dreadful child. This is what is seen."[2] Everyone in the town is happy because the glazier is continuing to be employed because of the boy's negligence. However, Bastiat calls the people misguided. They are missing the unseen. If the window had not been broken, then the money that was paid to the glazier could have been spent on something else that was productive. Someone else could have used the money that was used to fix the window to create some new product instead. The people are not seeing the opportunity cost of the money that was used to fix the destruction.

The Broken Window Fallacy is a fallacy exactly because it seems logical at first, but it is actually pure nonsense. By taking the fallacy to the extreme, it becomes more obvious, as it implies that we should destroy everything so that people will have jobs. Hurricanes and earthquakes would be major blessings because the massive destruction would require rebuilding, which means opportunities for employment. The desolation of war would be a stroke of good fortune. This line of thinking is ridiculous because the destruction of everything leads to poverty, not wealth. *Production* creates wealth, not *destruction*. The opportunity cost of rebuilding or repairing broken things means that money, time, and resources can't be devoted to innovating and creating new things that make human life better.

Government policies are often the source of negative unintended consequences. For example, many Americans see that the country's producers are threatened by competition from foreign imports, like steel. They call on the government to fix the problem, so then the government passes a tariff on imported steel. However, the most common result is that the price of steel rises, and all industries that use steel in their products have to raise prices for consumers. Additionally, international politics is often a tit-for-tat game, and foreigners choose to buy fewer American goods or put tariffs on American goods as retaliation. This then harms American producers as well. So, tariffs usually end up benefiting a very small group—one privileged group of domestic producers—while hurting nearly everyone else, including consumers and other producers.

There are other common examples of negative unintended consequences from public policies. For example, city officials often see that apartment rents are rising, which puts decent, affordable housing out of the reach of less wealthy people. In past decades, as well as more recently, many municipal officials have called for rent controls to stop the rising price of housing. Despite what is a noble intention, rent controls can and often do have perverse effects that harm the very people they are meant to help. The economics literature has shown a tendency for landlords to skimp on the upkeep and repair of apartments under rent control. Also, in the long run, developers build fewer rental units and focus largely on luxury housing because the rent control policy limits their ability to make a profit, or in many cases, break even.

The War on Drugs is another example of a public policy that has led to many negative outcomes.[3] By toughening the enforcement of drug laws, the prices of drugs on the black market rise, which can lead addicts to engage in riskier and more violent behavior. Black markets also induce organized crime activities so that people can share the risks of engaging in illegal activity. Incarceration rates increase, which breaks up more families, leading to more social problems down the line, especially for minority communities. The potency of drugs also increases, leading to more overdose deaths. Aggressive drug policies also lead to the erosion of civil liberties, such as stop-and-frisk, no-knock raids, and asset seizures without due process.

Although many of the most common unintended consequences are negative, some unintended consequences are actually beneficial, both for individuals and society. Perhaps the most important beneficial unintended consequence is a spontaneous order, which is a pattern or system that emerges from human action, but it was not purposefully designed by anyone. In other words, it is an orderly pattern that arises, even though no one is consciously directing it.[4]

Adam Smith recognized spontaneous orders in *The Wealth of Nations*, which he famously called "the invisible hand." First, Smith writes that people trade with one another, not necessarily because they love their neighbors, but because they are concerned about their own livelihoods: "It is not from the benevolence of the butcher, the brewer, or the baker that we expect our dinner, but from their regard to their own self-interest. We address ourselves not to their humanity but to their self-love, and never talk to them of our own necessities, but of their advantages."[5] Later, Smith discusses that as people work for their own self-interest, they provide their neighbors with the goods and services that they need to make their lives better. As Smith writes, a man "intends only his own gain, and he is in this, as in many other cases, led by an invisible hand to promote an end which was no part of his intention."[6]

The spontaneous order of the market is a "marvel," as Hayek called it, because the goods and services that we want come to us at roughly the right time, in roughly the right place, in roughly the right quantities and qualities.[7] In the eighteenth century, Adam Smith saw the marvel of the market's spontaneous order with the creation of a "common woolen coat,"[8] and twentieth-century economist Leonard E. Read saw the marvel in the creation of a pencil.[9] Smith and Read recognized that a simple wool coat and simple pencil are incredibly complicated to make. If you want to make a wool coat, you need to raise sheep, shear the sheep, scour the wool to remove impurities, card the wool to remove any remaining impurities, comb the cleaned wool, spin the wool in yarn or thread, weave the wool threads into a cloth, dye the wool cloth, sew the cloth into a coat, and transport the new coat to a store. Any one person would have a very difficult time learning and doing any one of these steps. However, these steps are just the tip of the iceberg. Someone must make the shears. Someone must make the buttons for the coat. Someone must do the accounting for the retailer who sells the coat. Someone must grow the food to feed each and every one of the people in the process.

The same goes for making a pencil, which requires rubber for the eraser, graphite for the lead, wood for the body of the pencil, and metal for the ferrule (the metal piece at the top of the pencil). Rubber must come from rubber trees in tropical locations, and people must extract the raw rubber, transport it, refine it, and transform it into pencil erasers. Graphite ore must be mined, transported, refined, and manufactured into pencil lead. Wood must be harvested, transported, and carved into a pencil shape. The metal for the ferrule must be mined, transported, smelted, and manufactured. Then, someone must take all these pieces and combine them into a workable pencil. Thinking about the inputs into each and every one of these steps involves possibly thousands of people. In the end, a total of millions, if not tens of millions, of people are directly or indirectly involved in the creation of a single pencil. The spontaneous order of the market is the marvel that allows this incredibly complex process to happen, even though there is no single person guiding the process. The order emerges from the self-interested actions of everyone who is simply looking out for their livelihoods.

The spontaneous order of the market is incredibly helpful because it reduces the time and energy that any one person needs to invest to live a better life. The creator of the YouTube channel "How To Make Everything" chose to make a chicken sandwich from scratch, so he grew his own vegetables, made his own salt from ocean water, milked a cow to make cheese, ground his own flour from the wheat he grew, collected his own honey, and raised and slaughtered his own chicken. The whole process took about six months and cost him about $1,500. After taking his first bite, a friend asked how the chicken sandwich tasted, and he responded, "It's not bad... That's about it. It's not bad. Six months of my life for 'not bad.'"[10] Even the arduous journey of making a chicken sandwich from scratch wasn't truly from scratch. He took a commercial flight to the ocean to collect salt water to boil out the salt. The process would likely have taken much longer and have been much more expensive if he had to either walk to the ocean or create his own plate from scratch. He didn't mine his own iron ore to make the knife that he used to cut the sandwich. This chicken sandwich shows how the spontaneous cooperation of millions of people across the globe allows us to live better and with fewer costs than if we tried to do everything on our own.

In sum, human life is characterized by unintended consequences. Selfless intentions can sometimes lead to damaging outcomes, and self-interested intentions can often lead to socially beneficial results. There may be many cases in which the proverb that "the road to hell is paved with good intentions" is true. If well-meaning policies lead to undesirable results, then we need to re-evaluate how the policies are set up. Additionally, spontaneous order, as an unintended consequence, is a "goose that lays golden eggs," so we should understand how spontaneous orders arise so that we don't accidentally kill the goose.

9

Public Choice Theory

All human action takes place in two "spheres"—the public and the private. The public sphere involves government action and includes voters, elected politicians, appointed bureaucrats, judges, and so on. The core characteristic of the public sphere is *coercion*, meaning that decisions are not optional. Anyone under a government is obliged to keep the law, or they will be forced to keep it.

For example, if you don't like paying income taxes, you can choose not to. However, the democratically created law requires you to pay income taxes, so government bodies, like the Internal Revenue Service (IRS) for the US federal government, will write you a letter telling you that you must pay your income taxes, and they may place a late penalty on you for missing the payment deadline. If you still choose not to pay your taxes, the IRS can put a tax lien against you, seize your property, or in extreme cases, pursue criminal charges. Although most tax issues go through civil processes—liens and levies—rather than criminal prosecution, neither civil nor criminal processes are optional. If you attempt to resist the actions of government officials, they can arrest you. If you resist arrest, they can use violence against you to subdue you. If you resist even more, in the most extreme circumstances, they may use lethal force against you.

This is what coercion means. Governments enforce laws, regulations, and other policies using force or the threat of force. Only government entities or institutions have the legal authority or power to compel compliance.

Depending on your moral framework, this power might be perfectly justified in some cases, but not others. For example, arresting people suspected of committing murder is widely viewed as a legitimate use of compulsion.

On the other side, the private sphere contains all action outside of government, which includes markets, companies, religious groups, nonprofits, nongovernmental organizations, clubs, civil associations, mutual aid societies, and so on. The private sphere is characterized by freedom of contract and voluntary entry-and-exit. In the market, you aren't compelled to buy any product from a particular company. You aren't forced to join a particular club or church. If you don't like what a particular company is selling, you can take your business elsewhere. If a club or church no longer fits your preferences, you can choose to go to another one that suits you better. Of course, action in the private sphere may involve difficult tradeoffs, such as choosing to leave a toxic workplace or leaving a close-knit religious community, but the presence of difficult choices does not necessarily imply coercion. When the tradeoffs become especially extreme, only then would action in the private sphere meet the threshold of coercion.

Traditionally, economics as a field has focused on the private sphere, especially markets. However, the tools of economics can also be used to understand the public sphere, which is called "public choice economics," "public choice theory," or simply just "public choice." By applying the assumptions, theories, and methods of economics to the realm of government, we can gain important insights about how the world works.

One of the core ideas within public choice economics is behavioral symmetry, which means that we assume that human nature is the same in both the private sphere and the public sphere. These assumptions go back to some of the previous tools discussed in this book. All people, whether they are in the private or public sphere, are assumed to be rational and respond to their incentives and constraints. They also weigh the marginal costs and marginal benefits of any decision. The reason that we see different behavior in the private and public sphere is not because people are greedy and self-interested in the market, but then they transform into selfless public servants when they enter a government job. Instead, institutions will affect the incentives, constraints, costs, and benefits that people experience.

Nobel Laureate James M. Buchanan, who was one of the main founders of the public choice approach, referred to it as "politics without romance" because he wanted a more realistic way of analyzing human behavior as it relates to government.[1] In contrast to the public choice theory of government action, a "romantic" view is public interest theory, which assumes that policymakers work solely for the betterment of society. However, public choice economics tries to make the analysis more realistic by assuming that governments are not composed of magnanimous statesmen who work for the common good. Instead, they are made up of normal people who respond to incentives and constraints, like anyone else.

All people, including politicians and bureaucrats, are guided by their own self-interest. However, self-interest does not necessarily mean nefarious selfishness. Many socially beneficial policies can come from government officials, but those policies incidentally align with the official's self-interest. There are also many cases in which government officials' self-interest leads them to engage in activities that benefit a small minority at the expense of a larger majority, leading to social waste. By removing any "romantic" notions about how governments work, we can analyze how they actually do work. Of course, if we think about how governments *should* work, that is a very separate question, and we should have high aspirations for creating good governments.

Public choice economics examines a wide variety of people in government, including voters, special interest groups, politicians, bureaucrats, and judges. Starting off with voters, people cast ballots for their government leaders in democratic societies, who make laws and determine how to spend money. But there is a bigger question involved—Why do people even bother to vote in the first place if their vote is highly unlikely to determine an election? As mentioned in Chapter 1, public choice economists explain this apparent paradox through expressive voting.[2] People choose to vote because they want to express a feeling or sentiment. For some people, they feel good fulfilling their civic duty. For other people, voting is like cheering for your favorite sports team. Sitting at home and cheering for your team from your couch doesn't actually help your team win, but it makes you happy to be part of the fun.

The "why" of voting can be explained by people wanting to express something, even if they know their individual vote doesn't influence the outcomes, but a more interesting question about voting is "how" people vote. When we look at how much voters know, the vast majority of them know next to nothing. To be fair, most American voters could probably name the president, their senators, and maybe their representative. However, they don't really know other relevant people or ideas that are important to policymaking. Most voters don't know much about economic policies, such as the effects of tariffs on prices or the source of inflation. They don't know much about environmental policies or the tradeoffs involved with implementing them. Most people know nearly nothing about foreign policy, especially since they can't locate most countries on a map, tell anything about the history of other countries, or know other countries' allies and enemies. People often have strong feelings about social policies, such as welfare, illegal drugs, and immigration, but they can't articulate what the root causes of these problems are, and they can't evaluate the relevant ways to effectively address them.[3]

What can explain the astounding ignorance of voters? Public choice offers at least two answers. The first is called "rational ignorance." Unless your job is a political researcher or policy analyst, then you have to give up a pretty significant amount of time, money, and effort to become relatively well informed about candidates and policies. When people make important personal decisions that directly affect their lives and livelihoods, they have a strong incentive to gather and process a relatively large amount of information—until the marginal cost of gathering more information is greater than the marginal benefit.

For example, let's say you're buying a new car. You're likely to do some research on different makes and models of cars, and you're likely to shop around at different places to test-drive some before you commit to buying one. However, voting is different. Since people realize that their individual vote is very unlikely to be the decisive one in an election, they can rationally choose to be ill-informed. In other words, people have a very weak incentive to gather and process relevant information when voting, so they rationally remain in the dark about most public policies and their implications.

To make the matter of rational ignorance more complicated, "rational irrationality" takes ill-informed voting to a new level. When we examine the data of what voters know, they hold systematically biased, low-information beliefs, but they hold these beliefs with a high degree of certainty. Since voters individually bear a low cost for their mistaken or misguided beliefs, they can indulge beliefs that don't align with reality. For example, you might mistakenly believe that if you flap your arms fast enough, you can fly. However, if you decide to test your theory by jumping off a skyscraper, you will pay a very high price. Voting is different. If you believe something wrong, or even crazy, and you vote based on those beliefs, there is no clear discipline mechanism to correct you. The outcome of an election will happen, regardless of how a person individually votes, so that the voter can continue with completely mistaken beliefs. In short, people rationally choose to continue believing biased and misinformed things because the feedback mechanisms of the voting process allow them to.[4]

Another set of important actors in politics is special interest groups, which are any organizations that seek to receive special advantages through political lobbying. The term "special interest group" often has a negative connotation, but it doesn't necessarily have to be negative. In some cases, a special interest group's mission might align with your moral worldview, such as groups that advocate for the environment and civil rights. However, other groups are more self-serving, and they look for advantages that protect their interests at the expense of other people. Recall that the most fundamental characteristic of a government is the power of coercion. Many special interest groups want to use that power for their benefit, so they try to persuade government officials to use that power on their behalf. All sorts of industries have formed a variety of special interest groups, including those in pharmaceuticals, insurance, technology, fossil fuels, renewable energy sources, education, real estate, defense, aerospace, and many others.

The process of special interest groups looking for favors and privileges through the coercive power of government is called "rent-seeking." A rent, in this case, is not the same thing as the money you pay to a landlord. Instead, rent is a term in economics that means a payment for the use of any resource, over and above its opportunity cost. So, rent-seeking is the expenditure of resources in an effort to contrive exclusive privileges. These privileges take many forms, such as legally granted monopolies, subsidies, loan guarantees, protection from foreign competitors, bailouts (or the promise of bailouts), favorable tax treatment, regulations that limit domestic competition, mandates to purchase particular products, and raising the costs of a rival.[5]

The problem with rent-seeking is that it is socially wasteful. Public choice views politics as a form of exchange among rationally self-interested individuals, and while market exchanges create new value, political exchanges only redistribute resources. Special interest groups try to influence politicians and bureaucrats to redistribute wealth and resources to favor themselves at the expense of other people. Politicians and bureaucrats cooperate with special interest groups in a quid pro quo way because these groups can influence voters with re-election, assist in the concentration of political power, or facilitate the expansion of bureaucratic budgets.[6] Normally, redistribution is a zero-sum process because one side's gain is another side's loss. For example, if I take $10 from Mary and give that $10 to Barbara, the total of social wealth is the same. But the government doesn't have unlimited money, so special interest groups compete among themselves for privileges, leading to negative-sum outcomes.[7]

From a social perspective, rent-seeking is wasteful because the total amount of wealth that all the special interest groups spend to get a single government-granted privilege is greater than the amount of the privilege itself. Additionally, redistributing wealth is not costless because of bureaucratic overheads, meaning that wealth has to be spent on the people who actually do the redistributing. On top of that, an "arms race" occurs between people who want redistributed wealth and people who don't want their wealth taken away, so both sides spend resources to secure their already existing wealth, making society worse off.[8] The processes of rent-seeking and the granting of government privileges show us that government decisions have externalities that spill over onto other people who they weren't meant to affect, just like externalities with market actions.

If the processes of rent-seeking and government privileges are socially harmful, then why do we see so much of it, and why don't we stop it? Public choice explains this through the logic of "concentrated benefits and dispersed costs." Usually, these government-granted privileges are concentrated on a small number of people, like those in a special interest group, but the costs are spread over a relatively large number of people, such as taxpayers. In many cases, special interest groups can reap large benefits from these privileges, such as millions or billions of dollars. But individual taxpayers are each forced to pay a relatively small cost, meaning that it is not worth it to fight back. In other words, taxpayers have a weak incentive to oppose policies that grant privileges, but the recipients of these privileges have a strong incentive to perpetuate such policies.

One of the classic examples of this logic is the sugar industry in the United States. Sugar cane is grown in Florida, Louisiana, Texas, and Hawaii, and American sugar cane farmers don't like competition from farmers in other countries that grow it more cheaply. The domestic sugar cane farmers persuaded the federal government to impose a tariff on foreign-produced sugar, making it more expensive, meaning that domestic consumers are more likely to buy American-grown sugar.[9]

Thanks to the tariffs, domestic sugar cane farmers collectively make millions of more dollars each year than they would if the tariff didn't exist. However, consumers who live in the United States each individually pay only a few dollars more per year than they otherwise would if the tariffs weren't there. The domestic sugar growers have a very strong incentive to keep the tariff going because they benefit hugely from it, so they put constant pressure on policymakers. However, it's not worth the time, effort, and money for any individual sugar consumer to fight back against a policy that harms them because the cost is relatively low. Even if a consumer took the time to write letters to their representative and senators, asking them to vote to repeal the tariff, the time to write the letters has an opportunity cost, and it's likely that the time to write the letter has a higher value than the additional money the consumer spends on cane sugar in a year. These government-granted privileges are especially long-lasting because the small number of beneficiaries gain much more individually than the larger number of people who are harmed to a relatively small degree.

At first glance, it may seem unrealistic that so much bald-faced rent-seeking happens in real life, but rent-seeking happens in a more complex way that hides the naked self-interest. Usually there are multiple groups calling for public policies, some of which want the policies for moral reasons, and others who want them for self-interested reasons. This is called a "Bootleggers and Baptists" relationship.[10] The "Bootleggers and Baptists" phrase is an analogy for two seemingly unrelated groups who have different interests and preferences, yet they work together for the same policy. In other words, special interest groups often seem to have strange bedfellows.

In the analogy, many Baptist churches in the southern United States would petition their local city or county governments to ban the sale of alcohol on Sunday for religious reasons. They saw alcohol as defiling the Lord's holy day, meaning that they had a moral reason for pushing for a regulation. Behind the scenes, bootleggers—people who sell alcohol illegally on the black market—

also supported the Baptists' call to eliminate the sale of alcohol on Sundays because that would limit the bootleggers' competition, and they could sell more alcohol on the black market. Sometimes, bootleggers masquerade as Baptists, using moral language to justify a public policy, but their real motives are for personal gain at the expense of others. There are many examples of "Bootlegger and Baptist" relationships in nearly all areas of policymaking, such as healthcare, energy, national defense, and many others.

As Bruce Yandle—the economist who coined the Bootlegger-and-Baptist theory—and his coauthor Adam C. Smith have argued,

> Bootleggers and Baptists make their influence felt as both legislators and regulatory agencies are developing regulations. Common Bootlegger goals include raising the cost of competitors, gaining subsidies, cartelizing industries (in whole or in part), and building protective regulatory walls around their sectors of the economy. The more new rules are issued, the more opportunities Bootleggers have to benefit—and the evidence suggests the stakes are only getting higher: the number of economically significant regulations, those whose effects on the economy are estimated at $100 million or more for each year the rules are enforced, is growing steadily.[11]

In the public choice approach, politicians—meaning people who are elected to office—rationally respond to their incentives and constraints, like all other people. First and foremost, politicians are primarily concerned with election or reelection. They may have grand visions of how they want to improve the world, but they cannot implement those visions unless they are actually in office.

Once politicians get into office, they need to act strategically to accomplish their goals. Most politicians are legislators, meaning they are members of congresses or parliaments that make laws and decide on spending. Legislators must act strategically to ensure that they are reelected by keeping a majority of their constituents happy. Two forms of strategic action in a legislature are important to highlight: logrolling and pork barrel spending. "Logrolling" is the term used to describe vote-trading among different legislators. In other words, legislators tell one another, "I'll vote for your bill if you vote for mine." By exchanging votes with one another, legislators can increase their chances that their preferred policies are passed.

Pork barrel spending refers to allocating government funds to local projects for a specific constituency or region, rather than the country as a whole. Legislators have a strong incentive to support programs and policies that provide benefits to the voters in their home district, no matter how irresponsible or wasteful the programs or policies are from a national perspective. Voters generally reward legislators who bring benefits to their district, such as public works projects, agricultural subsidies, and military bases, but when all legislators engage in this behavior, spending is much higher across the country as a whole, and taxpayers in one place are heavily cross-subsidizing taxpayers in many other places.

In recent years, one of the most blatant examples of pork barrel spending was Alaska's infamous "bridge to nowhere."[12] The proposed Gravina Island Bridge received $400 million in federal funding to connect Ketchikan, with a population of 8,000, and Gravina Island, with a population of 50. National outrage killed the project, leading to some restrictions on "earmarking" money in Congress for particular projects. The changes granted more discretion to agencies to spend government funding.

After legislators make laws and decide on spending, bureaucrats are the government employees who implement and enforce those decisions. Unlike politicians, bureaucrats are appointed or hired to an office rather than being elected. To implement the laws that a legislature has created, bureaucrats use their expertise to make specific regulations based on the laws. Most legislators don't have specific expertise in many fields, but laws are made on every topic imaginable, so they need to rely on specialists to make sure that the intentions behind laws are actually fulfilled. In the modern world, most legislatures delegate broad discretion to bureaucracies to make the nitty-gritty rules that govern daily life.

Like all other people, bureaucrats are rationally self-interested and respond to their incentives and constraints, even though they don't have incentives to win office. One of the major incentives that bureaucrats face is to maximize their budgets.[13] Legislatures decide on how much public money will be allocated to a bureaucracy, and bureaucrats, like anyone else, would rather have more than less. This means that bureaucrats have a weak incentive to spend money prudently, and they have a relatively strong incentive to slightly overspend each year. By overspending, bureaucrats can show to a legislature that they need a bigger budget. If bureaucrats were to work as cost-efficiently as possible, they might spend less than their allocated budget, but the legislators would see this and trim down the budget the next year, which may make it harder for bureaucrats to do their job. The end result is that the scale and scope of government tends to grow over time, and few examples exist of governments actually shrinking in size and scope.[14]

Besides budget maximization, bureaucrats face a number of other incentives. Bureaucracies are organized like a pyramid, with one decision-maker at the top, a few second-tier people beneath that, more third-tier people beneath that, and so on. The higher a bureaucrat climbs up the pyramid, the more money and decision-making power they have. This means that bureaucrats have a strong incentive to do what it takes to climb up the hierarchical ladder, which might involve flattering your bosses, gaining the loyalty of your underlings, working to increase the bureau's budget, etc.[15] By working up the ladder, bureaucrats gain greater prestige for themselves, gain a large degree of administrative discretion, and have a wider ability to implement their own preferred policies and worldviews. Bureaucrats don't have free reign, however. They still have constraints, like congressional oversight, hierarchical oversight within a bureaucracy, competition among different bureaucracies in the same government, and other legal constraints. So, even though discretion is fairly broad within a bureaucracy, it is by no means unlimited.

Judges are human, too, meaning they are rationally self-interested. The public choice approach assumes that judges are motivated by the desire for power and prestige, so they want to make decisions that will not be overturned or too heavily scrutinized. By proving that they make respectable and prudent decisions, they can gain advancement through the hierarchy of courts, all the way to a country's or state's supreme court. By carefully using their position, they can help forward their personal worldview and ideology, as long as enough people continue to see the decisions as legitimate. In some locations, judges are elected, which means they face a new set of incentives similar to other politicians, which can make judicial decisions much more partisan than judicial decisions in places with appointed judges.

It's also important to note that political processes generate different types of knowledge than market processes. Recall that markets have prices that allow us to aggregate and communicate knowledge that is dispersed. Markets also have the feedback mechanism of profit and loss, which disciplines the behavior of entrepreneurs so that they devote resources to goods that consumers actually want. Political processes also have feedback mechanisms, like voting, but those sources of feedback are often less clear and don't necessarily aggregate and communicate knowledge as well. For example, voting is not always a clear feedback mechanism for politicians because voters are rationally ignorant, if not rationally irrational, so there is often a disconnect between what voters want and what they vote for. Voting as a feedback mechanism is often ineffective because voters must choose candidates who represent a bundle of policies, which forces single-issue voters to support a candidate who aligns with their priority issue but may advocate for many other policies they strongly oppose. Additionally, mistakes or inefficiencies in government policies can be harder to identify and correct because politicians and bureaucrats do not have the equivalent of profit-and-loss signals.

In summary, public choice economics leaves us with several lessons. Focusing on getting the "right" people into power is a shallow solution to a complex problem. Obviously, "better" people in power can help address some social problems, but it is not a permanent solution. Instead, public choice directs our attention to the institutions that people operate in, and how those institutions shape the incentives and constraints that people face. Institutional problems demand institutional solutions. "Better" people are not likely to fix a problem in the long term because placing good people in an institutional environment with perverse incentives or weak constraints will likely induce them to act poorly.

10

Polycentric Governance

Nobel Laureate Elinor Ostrom and her husband, political scientist Vincent Ostrom, developed a new way of thinking about and addressing social problems, which has been called the Bloomington school of political economy (named so because they were based at Indiana University Bloomington). The Bloomington approach is centered around one main question: Is it possible for people to solve complex collective-action problems without top-down, centralized, government-based solutions? As discussed earlier, there is a distinction between normative and positive analysis, and the Ostroms' positive analysis started with a foundational normative commitment to democratic self-governance. They were afraid of tyranny in all its potential forms, so they wanted to develop a science of self-governance, allowing people to devise their own rules for overcoming the various social problems they faced.

The Ostroms' normative worldview distinguished between governing *over* others versus governing *with* others. By governing *over* others, they saw politicians and bureaucrats as foisting policies on people who had little direct say in the way these policies were formed, and they had little ability to push back against policies that didn't fit the prevailing local preferences. Instead, the Ostroms wanted to understand how people could govern *with* one another. Governing with one another can take many forms, and public administration from government may be necessary in many cases, but then the points of concern are not about whether government will do something or not. The real questions are as follows: What level of government is most appropriate for any problem? What are the rules that will constrain government action? How much creativity and adaptation will be allowed in public policy? What role will markets and civil society also play in solving a social problem? The Ostroms' vision is that a self-governing, democratic society should allow free and responsible individuals to find ways of solving their social problems in ways that avoid conflict and facilitate cooperation.[1]

The Ostroms' work follows in the tradition of Alexis de Tocqueville, who was a nineteenth-century French statesmen and scholar. One of Tocqueville's most famous works is his book from the 1830s called *Democracy in America*, which is an account of his observations as he traveled around the United States. In *Democracy in America*, Tocqueville analyzed how American society functioned and marveled at the robustness of American civic life. On the frontier where the federal and state governments had very little presence or power, people formed associations that provided their communities with *governance*, even when there wasn't really any effective *government* to speak of. In this case, governance refers to orderliness that arises through the creation and enforcement of rules for social cooperation. Governance can come about in many cases without government. If you belong to a club, organization, or church, it is very likely that those groups have rules, which means you have experienced a form of governance that isn't tied to government. The Ostroms' goal was to figure out how successful governance actually arises.

One of the major insights that comes from the Ostroms and the Bloomington approach is that there are no strict dichotomies in governance. In other words, the options are not just between markets or states. There is also civil society. And markets, governments, and civil society have many different levels and facets. In markets, some firms are very large, like international corporations, and many are very small, like a mom-and-pop shop on your local Main Street. In governments, power is often divided across many levels and locations. For example, in the United States, the federal government has power for countrywide issues, state governments can make rules for themselves, and even counties and cities within states have their own rules and regulations. In civil society, many different groups exist, with differing degrees of organization and permanence. Churches, synagogues, mosques, clubs, civic associations, mutual aid societies, and other nonprofit organizations are just some of the different types of civil society groups that exist.

The realization that there are no strict dichotomies pushed the Ostroms to study how polycentric systems can address complex social problems. "Polycentric" means "many centers," as opposed to "monocentric," or one center. In a polycentric system, there are multiple decision-makers who have some degree of independence, but they are also interdependent. In other words, the decision-makers overlap one another and are nested within one another to varying degrees. One decision-maker doesn't dominate all the others. Some decision-making centers may have more power than others, but that does not necessarily undermine the system.

Many examples of polycentric systems exist. In terms of government, federalist systems divide power vertically into a national government, subnational governments (like states or provinces), and even smaller levels, like counties and cities. At each of these levels, power is divided horizontally among different branches, including the legislative, executive, and judicial. Different forms of federalist systems exist in the United States, Canada, Mexico, Brazil, Argentina, Germany, Switzerland, South Africa, India, and Australia, among others. A monocentric government, however, would be a pure dictatorship in which one person makes all the decisions, and others must comply.

In the private sphere, polycentricity abounds. Markets consist of many different individuals who come together in companies that compete against one another, but they also cooperate with one another. A monocentric market is simply a monopoly—one company controls an entire industry. Civil society has many organizations, associations, and nonprofits that act independently of one another, but sometimes they work together. The scientific community is composed of many researchers that act independently on their research, but sometimes they collaborate. If they disagree with other research that has been put out, they can push back against it with their own research. If the scientific community were monocentric, then it would function much like an Orwellian "Ministry of Truth" that singularly decides what is "correct" and what is not.

The growing literature on polycentricity examines how dispersed decision-making helps solve complex social problems. The body of research on polycentric systems highlights at least six advantages that polycentric systems have for coping with complex issues:[2]

1. Competition among different entities
2. Cooperation between different entities
3. Legitimacy and coproduction
4. Experimentation and learning
5. Resilience
6. Spontaneous order

When a system is polycentric, the different decision-makers compete against one another, which provides stronger incentives to make more socially productive choices than otherwise. The clearest example of competition is in markets. When one company has to compete against another, they have stronger incentives to make products of higher qualities and at lower prices. They also have an incentive to innovate and make new products that consumers will want to willingly buy. They also have a stronger incentive to cater to the desires of consumers because if they don't, their competitors will. If a company is a monopoly, they have relatively weak incentives to create newer and better products, and they purposefully keep prices higher than they would be in a competitive market. Relatedly, in civil society, people can "shop around" and find the clubs, churches, and nonprofits that best fit their preferences. If one of these groups doesn't meet the expectations of its patrons or members, then those patrons and members will choose to go somewhere else.

Competition also takes place in governments. If you don't like the policies or taxes in one state, you can choose to move to another state that better fits your preferences. Policymakers have an incentive to keep their constituents relatively happy because they could move away, taking their tax dollars with them. "Voting with your feet" is one way to keep policymakers accountable because policymakers in each jurisdiction have an incentive to entice new residents and investments to flow in.[3] Of course, moving to a new place is costly, and not everyone can afford to do it. However, the fact that many people can move away provides some discipline to government officials, and they will act with more discipline than a world in which taxpayers are captive and can never "vote with their feet." The worse the policies are, the more likely the people will move away, and the fewer taxpayers there are to fund a policymaker's ideas.

Besides competition, cooperation is an important aspect of polycentric systems because different entities can come together to solve problems, leading to a synergy in which outcomes are better than the sum of the parts. In markets, we see firms cooperating all the time. A common example is airline companies partnering with credit card companies in co-branded credit cards. Customers earn frequent flyer miles when they use their credit cards to make purchases, which benefits the customer, the airline, and the credit card company.[4] Another example was during the COVID-19 pandemic when Pfizer, one of the world's largest biopharmaceutical companies, partnered with BioNTech, a leader in mRNA research, to develop one of the major COVID-19 vaccines. The relationship was mutually beneficial for the two companies because BioNTech brought innovative technology, and Pfizer brought its vast resources of manufacturing and distribution. The synergy created by these two companies led to a successful and rapid vaccine rollout.[5]

We see cooperation often in governments too, especially in the form of interjurisdictional compacts. For example, the Great Lakes Compact involves the US states of Illinois, Indiana, Michigan, Minnesota, New York, Ohio, Pennsylvania, and Wisconsin, plus the Canadian provinces of Ontario and Quebec. This compact is a mutual agreement among all these political entities that manages the use and conservation of the Great Lakes' water resources.[6] Managing the Great Lakes is too big for one state or province, so allowing both the national government and the subnational governments to have a say ensures more effective governance.

The Colorado River Compact includes the US states of Arizona, California, Colorado, Nevada, New Mexico, Utah, and Wyoming. This agreement allocates the water rights of the Colorado River among these seven arid states.[7] It makes sense for these seven states to cooperate since it directly affects them, and one state's actions can affect another. However, the federal government doesn't necessarily need to be the level to make the decisions because the management of the Colorado River doesn't affect far-away states like Vermont or Alaska.

At a bigger scale, the European Union is a supra-national entity of twenty-seven sovereign countries that facilitates political and economic integration among the member states, which makes it easier to conduct economic activity and security policies.[8] Thus, cooperation between various entities means that different levels can come together when necessary to solve a problem at the correct scale.

For any system of governance to work effectively, the people who are governed have to see the rules as legitimate, and they must contribute to the monitoring and enforcement of the rules. In the simplest terms, a rule—whether formal or informal—is legitimate if the people being governed by the rule buy into it. If they don't see a rule as legitimate, then they will either ignore or evade it. In the most extreme cases, they might even try to subvert or undermine the rule if they see it as completely illegitimate. People legitimize rules when they feel like their voice has been heard, which means either (1) they were directly involved in the rule-making process, or (2) somebody they view as their legitimate representative was involved.

If a rule is created in a way that makes the people being governed see it as legitimate, then those being governed help with the monitoring and enforcement of those rules. This process is called coproduction, which is when the consumers of a good also help produce it.[9] A classic case of coproduction is education. A teacher can try to teach, but learning will only happen if the student is listening and engaged. The same thing happens with public safety. If public safety were completely dependent on police, then the world would be a much more dangerous place. Public safety involves police, but it depends on neighbors reporting suspicious and illegal behavior to police, and then ostracizing community members who violate the rules. Any governance system will work only when the people being governed also contribute to the monitoring and enforcement of rules.

One of the most important characteristics of a polycentric system is the experimentation and learning that happens in various locations and scales. One company can observe what other companies do, and they learn and adapt from the successes and failures. In government, policymakers can see what did and didn't work in other places. They can adapt the successful policies in other places to their unique circumstances, and they can choose to avoid the policies that failed. In civil society, different clubs, churches, associations, and nonprofits can try out different rules or organizational structures, and other groups can adjust their approaches based on what they see. Since polycentric systems have a relatively large degree of decentralization, smaller groups and lower levels can more easily access different types of knowledge, especially knowledge that is local, dispersed, inarticulate, and tacit. When people face "unknown unknowns," one of the best ways to learn is through a trial-and-error process. If there is only one or a few decision-making centers, the learning process is slow. However, when there are a relatively large number of decision-making centers, trial-and-error learning is happening at many scales and scopes, making the whole system a kind of laboratory.

Since decision-making is dispersed in a polycentric system, the whole system is more resilient to failures.[10] When decision-making is centralized, a failure of any kind is necessarily a system-wide one. However, if decision-making is decentralized to some degree, any failure will affect only part of the system, leaving the other parts of the system intact. For example, in government, if Texas makes an education policy that leads to worse outcomes for the students in public schools, that does not directly affect the students in Massachusetts. If the US Department of Education makes a poor policy, the states have a relatively large amount of decision-making discretion, and they can work around failures at the federal level. However, if the US Department of Education were completely in charge of all education policies across the country, then a failure at the federal level would inevitably affect all people in the country.

The benefit of a polycentric system is that checks and balances exist in many directions, with lower levels balancing failures at the higher levels, and vice versa.[11] The same principle applies to markets and civil society. If a monopoly were to fail, all consumers would suffer, but due to the competitive nature of most markets, other firms have an incentive to capitalize on the failings of other companies by providing better goods at lower prices. Civic groups, like clubs, associations, nonprofits, etc., each provide checks and balances on one another. If one church fails to provide the theology or community services that a parishioner wants, then he can choose to go to another church. Civic groups don't like to lose members or donors, so they have an incentive to ensure that they cater to a wide variety of preferences, making the whole system more resilient to localized failures.

A final characteristic of polycentric systems is that socially desirable outcomes can emerge from the complex interactions of the different nodes in the system, but those outcomes don't have to be centrally planned. In markets, the goods and services that people want are produced and made available, but no person centrally directs how all goods should be produced or where they should be sold. Since markets are generally polycentric, a spontaneous order can arise that allows for human flourishing. Although the spontaneous order of the market is one of the clearest examples, the polycentric nature of governments and civil society can also lead to socially desirable outcomes.[12] For example, education policymakers in various locations might try different approaches, and through the processes of experimentation, learning, and adjustment, policymakers can create better educational policies over time. If decision-making were centrally planned, there would be less room for experimentation and learning, and adaptation would be slower.

The combination of many approaches at once can and often does lead to the emergence of workable solutions to complex problems. Even global-sized problems don't necessarily need to be globally governed. In fact, because of the sheer diversity of preferences and values around the world, attempts to govern social problems at the global level rarely work.[13] However, when lower levels of decision-makers try different approaches, the combined efforts lead to effective solutions. In short, complex social problems don't require top-down solutions, even though it may seem like they might at first glance. Bottom-up processes can and do solve large-scale, complex problems because regular on-the-ground people have relevant knowledge that is necessary to solve the problem. People also have stronger incentives to be involved in monitoring and enforcing social rules when those rules come from the bottom up.

PART TWO

APPLYING ECONOMIC TOOLS TO ENVIRONMENTAL PROBLEMS

11

Positive-Sum versus Negative-Sum Environmentalism

Now that we've assembled a toolbox of economic concepts, we can directly apply those tools to environmental problems. This chapter takes a broad perspective, highlighting three overarching ideas. First, institutions channel and shape behavior, which means that the root cause of all environmental issues is *institutional*. Second, people are *entrepreneurial*, and they can find innovative ways of addressing problems when they have the opportunity. Third, *positive-sum* interactions are more likely to effectively address environmental problems than zero- or negative-sum interactions. The distinctions among positive-, zero-, and negative-sum situations are crucial. Positive-sum games are those in which everyone involved wins, and interactions among people are mutually beneficial. The proverbial "pie" gets bigger, and there is more pie to go around. A zero-sum situation is one in which a person's gain is another person's loss. The pie remains the same size, and the more slices that one person takes, the fewer slices are left for other people. Negative-sum games are those in which the pie gets smaller, and people fight among themselves to lose the least amount—the "winner" isn't really a winner at all if you look at the big picture.

This chapter highlights contributions from several economists, including Nobel Laureate Elinor Ostrom, economist Julian Simon, and "free market environmentalism" (FME) movement founders Terry L. Anderson and Donald R. Leal. Ostrom, Simon, Anderson, and Leal, among many other economists, have studied ways in which certain institutions can channel entrepreneurship in positive-sum, socially productive, peaceful ways, rather than negative-sum, wasteful, conflict-ridden ways. The tools of economics can help scholars, policymakers, and environmentalists find win-win approaches for even the thorniest ecological problems.

As mentioned in Chapter 5, environmental problems almost always arise when private property rights are not well defined or well enforced. When a resource such as land or wildlife is owned by no one or owned collectively, it can fall into the "tragedy of the commons," leading to overexploitation and degradation. Another environmental problem is negative spillover effects—externalities—which harm people who had nothing to do with creating the problems. Air pollution or water pollution, at the core, are problems with property rights. Since establishing property rights to air or water is difficult and costly in many cases, people can let their waste products flow into waterways or be released into the air because there's not a clear liability for harm to someone's property.

When we encounter an environmental problem, we should ask ourselves why property rights are not well defined or well enforced. Perhaps it is too difficult or costly to establish property rights. This leaves us with three general categories of approaches. First, we can turn to government to create laws and regulations. Second, assuming that people can discover creative ways to establish new forms of property rights, entrepreneurial action and mutually beneficial trade can help to resolve problems. Third, civil society—including communities, private associations, clubs, religious groups, nonprofits, and mutual aid societies—can create their own rules to manage resources or cope with externalities. The choice between using government, markets, or civil society is rarely clear cut. Depending on the nature and scale of an environmental problem, there may be important roles for each one of these different spheres of human action. It's important to know the strengths and weaknesses of each type of approach so that we have a realistic assessment of how we might tackle real-world problems.

Many environmentalists assume that the default to solving ecological problems is government-created legislation and regulation. In other words, the assumption is that top-down government control is necessary to conserve species, preserve land, protect water resources, and combat climate change. Sometimes creating private property rights is not practical or possible, such as with property rights to air. If we want to control air pollution or greenhouse gas emissions, then we might need to turn to government to create laws and regulations. However, not all public policies are created equal. Good intentions do not necessarily mean that good outcomes will follow, which is why the concept of unintended consequences is crucial. Pragmatic environmentalists must be aware that government failure is a real and pressing problem. Additionally, simply saying that a government needs to address a problem doesn't tell us anything about *how* the government should do it. It also doesn't tell us about *which level* of government is best suited.

Many government policies create winners and losers, so they start off as zero-sum games. People end up fighting to avoid being the loser, which leads to conflict rather than cooperation. One of the many problems with conflict, especially political conflict, is that it often turns into a negative-sum game. To avoid being a loser in the political process, people use valuable resources like time, money, and effort, which could have been used to create value elsewhere. For instance, a company could spend its time, money, and effort by investing in innovation for more ecofriendly goods or efficient production techniques. Or, alternatively, that same company could spend millions of dollars lobbying for a regulation that blocks competitors. The money spent on lobbying didn't create new wealth or social benefits. In other words, the money spent on securing government-granted privileges made the overall "pie" smaller.

The public choice perspective in economics views legislators, bureaucrats, and judges as humans who also respond to incentives and constraints. Legislators, bureaucrats, and judges are not all-knowing, all-powerful, or perfectly benevolent. Despite their expertise, they can and do fail. As the founders of FME have argued, "Although government regulation has the potential for improving environmental quality and resource stewardship, the government-knows-best, command-and-control mentality requires assuming that centralized policy makers will accurately account for all costs and benefits and act to improve efficiency."[1]

Therefore, people calling for a government response to an environmental problem need to consider several points. Perhaps the incentives of legislators and bureaucrats sometimes lead them to protect special interest groups at the expense of other groups. Perhaps legislators and bureaucrats lack the relevant knowledge to solve complex environmental problems. Maybe it is difficult to foresee the potential unintended consequences that can arise from a proposed policy. It can also be difficult to assess the tradeoffs and opportunity costs that will accompany a proposed policy, especially when it comes to deciding who will bear those tradeoffs. Even if we think a policy might help address an environmental problem, the degree to which we should implement a public policy is not always straightforward. When we observe a "bad" environmental outcome, it may not be clear whether that outcome is the fault of a previous government policy, like subsidizing the use of fossil fuels or subsidizing water consumption.

One especially important insight from public choice economics is "Bootlegger and Baptist" relationships (see Chapter 9). Economists Bruce Yandle and Adam C. Smith have written on how and why these relationships arise in environmental policymaking. As they explain,

Fervent environmentalists make good Baptists. In their view, uncaring polluters are willing to trade the last landscape for another shot of positive cash flow. To prevent such rampant greed, public-spirited defenders of the Earth must expand and enforce environmental laws to punish firms that put private profit over the public interest. Then there are the Bootleggers, industrialists who prefer polluter profits to no profit at all. Just the right form of government regulation can be a dream come true for profit-loving Bootleggers. Although perhaps never fully comfortable in a room of self-righteous Baptists, Bootleggers still welcomed their religious counterparts when pious pleading can be parlayed into persistent profits.[2]

Giving a more concrete example, Yandle and Smith look at the rise and fall of the Kyoto Protocol, which was an international agreement to curb greenhouse gas emissions. The protocol was adopted in 1997, and entered into force in 2005, but the United States—the world's biggest emitter at the time—did not ratify it. Yandle and Smith write,

> When viewed through the public interest lens, the Kyoto plan looks a lot like an enlightened effort to save the planet from human harm. [...] Viewed through the Bootlegger/Baptist lens, Kyoto is about redistributing income from higher- to lower-income countries and creating a process that enables interest groups to build profitable cartels. Kyoto delivered a bright green invitation to a major-league pork cookout.[3]

In essence, Yandle and Smith argue that the behind-the-scenes deals to ratify the Kyoto Protocol in the United States were full of special interest groups hoping to benefit themselves at the expense of others. In other words, they were hoping to concentrate benefits on themselves, while dispersing the costs. The various stipulations in the Kyoto Protocol would benefit certain companies and countries through subsidies, favorable regulations, mandates, tax cuts, and raising rivals' costs, among other policies. Over time, however, the proposed policies that were based on the Kyoto Protocol's principles gained more enemies than supporters, and the pro-Kyoto coalitions lost many of their Bootleggers, while some of the anti-Kyoto groups lost their Baptists.[4]

As Yandle and Smith describe,

> Kyoto provides a prime illustration of how Bootlegger/Baptist coalitions form, transform, and ultimately dissolve. It also shows what happens when the forces of a major recession seriously erode the ability of politicians to dole out the rewards that hold those coalitions together. In a sense, the political market for Kyoto-based favors shut down. But it is not dead: new groups can always be formed when the time is right, after all. [...] As the economy recovers and new regulatory proposals surface, Bootlegger/Baptist coalitions are likely to prove renewable and sustainable.[5]

While this short recap cannot give all the details or nuances of the Kyoto Protocol, this example illustrates how seemingly beneficial policies might be undermined by Bootleggers who seek government-granted privileges. As Bootleggers engage in rent-seeking, resources are diverted from socially productive uses, making the overall situation negative sum.

Although government action often creates zero-sum and negative-sum scenarios, it can also be important for setting up positive-sum games by defining and enforcing property rights. Ideally, governments establish clear rules that define the ownership of something, specifying what constitutes property, who owns it, and what kinds of actions owners can take. These actions include how property can be used, leased, rented, sold, transferred, or split up. The clarity over the "what," "who," and "how" reduces the chances for disputes and uncertainty. With more certainty and stability, people can use their property for the things they value, which might include using it for economically productive purposes, or it might include keeping property unused for environmental conservation. Governments can also help make sure that owners can securely use their property by deterring and punishing violators. Property rights do not necessarily have to look like those in the United States or western Europe. They can take many different forms, according to local cultures, customs, and understandings. For property rights to be useful, everyone in a society must understand *what* the property is, *who* has the rights to make decisions, *how* the property can be used, and who will be *enforcing the rules* about any violations.

Once property rights are effectively established, markets and civil society can spark many creative and innovative ways to address environmental issues. The growing movement of free market environmentalism focuses on how property rights, markets, and civil society can lead to the discovery of positive-sum approaches for overcoming ecological problems. When most people hear the term "free market environmentalism," they might think it is an oxymoron or a dogmatic devotion to markets. For many environmentalists, they may ask themselves, "How can markets and environmentalism go together? Markets are the cause of environmental problems, so how could they be the solution?" FME is both a scholarly study of how markets and civil society can discover positive-sum approaches, but it is also a practice that people across the globe are engaging in. Scholars and practitioners are exploring how markets, civil society, and voluntary action can effectively address environmental problems without the need for centralized, top-down, coercive control.

The core idea of FME is comparative institutional analysis, meaning that environmentalists should compare the various means that are available to accomplish the ends that they seek. In some cases, markets or market-like regulations may accomplish the environmental goals more effectively or efficiently than top-down, command-and-control regulations. Legislation and regulations might have negative unintended consequences that undermine what environmentalists are trying to do. The people who are least fortunate may bear the brunt of the costs or tradeoffs associated with environmental policies, which may outweigh any benefits the policies yield.

The concept of FME forces us to rethink how we conceive of addressing environmental problems. We can take either an "engineering" approach or an "economic" approach. We might alternatively call this an "allocation" versus a "coordination" paradigm. The engineering or allocation approach is intuitive to most people because it focuses on how limited material inputs translate into limited outputs. In other words, an engineer has a set number of resources that have to be managed and allocated to reach a goal. An expert plans a system from the top down to address the problem. The economic or coordination approach, in contrast, focuses on processes of adaptation and discovery in a constantly changing world, especially when people disagree on ends and means. Unlike an engineer, economists recognize that there isn't a set number of tools or resources, which means that addressing a problem is a dynamic process as conditions change or new discoveries are made.[6] In other words, the economic or coordination paradigm involves bottom-up processes in which feedback mechanisms allow many people to adjust to one another and to external conditions.

Although there are many potential ways to address environmental problems, each way will involve its own set of costs and benefits. In other words, there are tradeoffs with any approach. Thus, any proposed solution to an environmental problem should evaluate (1) the incentives and constraints that individuals will face, (2) who will bear the costs and gain the benefits, (3) any potential unintended consequences, and (4) the ability to discover, aggregate, and communicate knowledge. By using these economic tools, environmentalists can become more informed about the source of environmental problems and suggest more pragmatic solutions. Ironically, the so-called dismal science of economics provides an important analytical framework for building a more hopeful, sustainable future.

By using private property rights and markets, the FME approach connects self-interest and resource stewardship. The key is property rights. Owners have no choice but to account for the costs and benefits of their actions. Poor decisions harm owners, and creative, innovative discoveries benefit owners. When people can bargain and trade their properties with one another, it can flow to higher-valued uses, which are often environmental uses.[7] When property rights cannot be enforced or transferred, conflicts arise. One person may assert that he is being harmed by the other, and clarity regarding property rights means that the other person must stop the harm and pay damages.[8]

The FME approach focuses on how cooperative, win-win situations are a practical, pragmatic way to solve environmental problems. The exchange of private property rights is almost always a cooperative, win-win situation because people would not willingly engage in an exchange if they thought they would be worse off. When private-property-based solutions aren't a viable option, public policy may be necessary, but there are better and worse policies. Policies that create a distinct group of "winners" and "losers" give the losers a strong incentive to fight back. Creating or reforming public policies to create cooperation and limit conflict will lead to more sustainable environmental outcomes.

Property rights aren't just defined and enforced out of the blue. The FME approach works largely because of government institutions that respect private property, uphold the rule of law, and limit the widespread practice of rent-seeking. However, the FME approach is always threatened by the fact that governments are made of fallible people who can be persuaded to act in socially unproductive ways. Government actors can choose to either strengthen and enforce property rights or weaken and undermine them. As the founders of the FME movement have written, governments can help "better define and enforce property rights and facilitate their transfer, [and] environmental markets can more effectively increase the value of environmental assets. On the other hand, if the government makes property rights less secure or inhibits transferability, transaction costs will rise and environmental markets will be less effective."[9]

The core logic of the FME approach is that property rights "create incentives for owners to know what they have, know what environmental goods they can produce, and know what demands there are for environmental resources. Environmental markets create information on all of these dimensions in the form of prices. Between these two institutions—property rights and markets—are environmental entrepreneurs who reduce the friction, which economists call transaction costs."[10] When natural conditions or human demands change, creative and alert individuals can discover and innovate. They might find new environmental resources, or they may recognize new opportunities for existing resources. They may discover demands for environmental goods. They might also find ways to get demanders of environmental goods to pay potential suppliers. When institutions get in the way of discovering or taking advantage of creative approaches, people can act as institutional entrepreneurs to improve upon property rights or the rules that govern them.[11]

The FME approach asks us to compare the realities of markets with the realities of politics. We need to add nuance to our thinking about causes, consequences, and potential solutions to environmental problems. We should rid ourselves of preconceived notions such as assuming "*the* government" must implement the solution. Governments are likely to be involved in addressing the largest and most pressing problems, but we need to look at the role of local and regional officials, nongovernmental organizations, and local groups of citizens.[12]

We should also look at the vast menu of policy options that are available. For instance, market-like regulations—including cap-and-trade systems—are often more effective at addressing problems than strict command-and-control measures. In fact, cap-and-trade creates a form of property rights as shares (or permits), and the total number of shares are "capped" by the government. Then, the government allocates those shares, and the recipients can trade them among themselves. Cap-and-trade systems allow people to come to mutually beneficial arrangements among themselves, even in an artificially contrived market. A command-and-control approach is often less effective because it locks in older technologies and provides a weaker incentive to make cleaner, more efficient technologies.

The problems with acid rain in the 1980s and 1990s were largely solved by a cap-and-trade scheme for sulfur emissions under the 1990 amendments to the US Clean Air Act.[13] Prior to 1990, the US federal government had a long history of setting command-and-control regulations under the Clean Air Act. However, this approach had many tradeoffs and negative unintended consequences. Command-and-control regulation did not provide a strong incentive for polluters to adopt cleaner technologies over time. These types of regulations also restricted compliance technologies and promoted "end-of-pipe" solutions, instead of pollution prevention and cleaner processes. Additionally, command-and-control regulations often sparked hostility and conflict between regulators and the regulated industry.[14]

In contrast to the top-down approach of command-and-control regulation, the market-like regulations of a cap-and-trade system have multiple benefits. Imagine there are two coal-fired power plants. Both are emitting black smog into the air, which mixes with the clouds, leading to acid rain. The rain then causes widespread devastation to plants, waterways, and aquatic life. The government could control the pollution by forcibly shutting down the power plants or commanding that they install certain technologies called "scrubbers" to reduce the emissions. However, imagine that one of the two power plants can reduce pollution at a lower cost than the other one. Perhaps the lower-cost polluter is newer, or maybe it uses a cleaner type of coal, or maybe it has another technology for reducing pollution that the other factory doesn't have, or maybe it can easily switch to a less polluting substitute. For the purposes of this thought experiment, the reason why the one factory can reduce its pollution at a lower cost isn't important. As long as one factory can reduce pollution more cheaply, then gains from trade are possible.

When looking at the whole system, we would want the newer, less costly power plant to reduce its emissions to a larger degree than the older, costlier one. In essence, we get more "bang for our buck" when the lower-cost reducer does more of the work. In other words, by having the lower-cost reducer do more of the work, we can do the same amount of work while wasting fewer other resources. It's important to note that there's no easy or apparent way for government officials to know which of the two power plants is the lower-cost one. Since there are countless ways of reducing pollution, there is no way that a single person could know all the tradeoffs. The knowledge about how to reduce pollution and the costs of reducing pollution is dispersed, and it's not easily centralized.

We can harness the power of entrepreneurs to discover, aggregate, and communicate this dispersed and tacit knowledge. Instead of commanding *how* exactly to reduce pollution, the government can facilitate a system that discovers the lowest-cost way of reducing pollution through a cap-and-trade system. By allowing the permits to be bought and sold between the two polluters, we can harness the self-interest of the power plant owners to reduce pollution. The lower-cost reducer is aware that he can reduce pollution at his power plant for a lower cost than he can get by selling the unused permits to the other power plant owner. The high-cost reducer sees that it is cheaper for him to buy the permits rather than to reduce his pollution.

By engaging in mutually beneficial exchange over the permits, the two power plants increase their profits at the lowest possible cost, while also reducing pollution. By spending less on reducing pollution, the power plant owners have more money and resources to devote to other important considerations, like improving energy efficiency or finding renewable energy sources. These new discoveries are likely to have positive spillover effects on the rest of society.

Of course, this example is highly simplified, but the underlying logic works for larger, real-world scenarios, and we've seen real evidence of this. Scholars were surprised at the effectiveness of the 1990 amendments to the US Clean

Air Act regarding sulfur emissions. In 2000, only ten years after the cap-and-trade system was implemented, sulfur dioxide emissions had been cut by over 50 percent of their 1980 levels. Additionally, compliance costs were much lower than scholars and government officials had initially anticipated.[15] The Regional Greenhouse Gas Initiative in the northeastern United States and EU Emissions Trading System have also been remarkably successful, outpacing many expert predictions.[16]

However, cap-and-trade systems have undeniable tradeoffs, which means they are no panacea. Under a cap-and-trade system, the overall pollution might be reduced, but particular locations may experience a concentration of pollution. This means that a cap-and-trade system may unintentionally harm people's well-being who live near higher-cost polluters. This is a difficult tradeoff, and there aren't any clearcut answers. We may need to be more creative and imaginative in how we approach these tradeoffs. Additionally, policymakers sometimes face political incentives to implement traditional regulations on top of cap-and-trade systems, which then undermines the function of market mechanisms, limiting the effectiveness and efficiency of the cap-and-trade system.[17] Another consideration is that when cap-and-trade systems are set up over large geographical areas, people may face relatively high transaction costs when attempting to engage in trading permits with one another. To make matters more complex, policymakers face knowledge problems when determining an appropriate cap, meaning they may set the cap too high or too low. If it's too high, the cap-and-trade system won't be especially effective at abating emissions. But, if they set the cap too low, the social costs will exceed the social benefits.

As professor of environmental law Ann E. Carlson highlights, "The general theoretical underpinning of cap-and-trade is to harness market forces to find the cheapest greenhouse gas emissions reductions by allowing emitters to trade allowances in search of the most efficient reductions."[18] Carlson continues, "when effectively designed, and when aimed at the right type of pollutants, cap-and-trade can deliver significant emissions reductions more cost-effectively than more traditional command-and-control or technology-based regulatory schemes. By the same token, experience has demonstrated potential difficulties with cap-and-trade systems."[19]

Cap-and-trade systems may be better than command-and-control systems on many margins, but that doesn't make them simple or easy to implement. Scholars and policymakers engage in institutional analysis to figure out the best rules and frameworks to achieve their environmental goals. An institutional analysis of a cap-and-trade system should consider a number of details: the kinds of pollution sources to be included; how share or permits will be distributed; what the scale, scope, and timeline of the trading system will be; who can buy or sell the permits; how market manipulation will be avoided; how the system will adapt to new information or unforeseen situations; and so on. These considerations leave a broad range of possibilities, meaning that a cap-and-trade system could take a countless number of forms. Some systems will be more efficient, effective, or adaptable than others. The big picture is that we often need multiple approaches to cap-and-trade systems so that we can learn from the successes and failures in each one and adjust based on what we learn.

Voluntary, positive-sum action doesn't just take place in markets or market-like regulations. In many cases, civil society—made of communities, private associations, and nonprofits—allows people to creatively discover and implement win-win scenarios that facilitate beneficial environmental outcomes. Some of these groups try to persuade other people to change their preferences or behavior. Others are more directly involved, buying land for conservation or financing eco-friendly technologies. Sometimes these groups discover new institutions or arrangements for producing beneficial outcomes.

One clear example of a civil society group bolstering pro-environmental results happened in 1995, when the US federal government chose to reintroduce wolves into Yellowstone National Park and the wilderness areas of central Idaho. Across the region, many ranchers and farmers were upset by the government's decision because wolves were likely to kill their livestock, meaning it would harm their livelihoods. However, Hank Fisher of the nonprofit organization Defenders of Wildlife wanted to ensure that the wolf reintroduction was successful. Fisher and Defenders of Wildlife chose to rethink the way in which they could help the success of the reintroduction. Fisher saw that it would be possible to use incentives to change the way that ranchers and farmers perceived wolves.

In essence, Defenders of Wildlife created a program that would pay ranchers the market value of any livestock that was killed by the newly reintroduced wolves. By implicitly granting the property right to the ranchers and farmers, Defenders of Wildlife struck a voluntary bargain to overcome the costs that wolf reintroduction would impose on unwilling people. As a nonprofit organization, Defenders of Wildlife raised funds through donations from people who were eager to see wolves return to the Rockies. The organization also commissioned an artist in Montana to create posters, depicting the return of wolves to Yellowstone. By selling the posters, Defenders of Wildlife generated more than $50,000 for the wolf compensation fund. The combination of outside donations and the revenue generated by the posters provided the resources that were necessary to facilitate a successful reintroduction. During the entire recovery process of the wolves in the northern Rockies, Defenders of Wildlife paid more than $1.1 million to ranchers and farmers who lost animals to wolves, compensating them for their losses and limiting the incentive for them to kill the newly reintroduced species.[20]

The experience of reintroducing wolves into the northern Rockies shows that people can cooperate when they are given the freedom to think and act creatively. As legal scholar Robert Ellickson explained, "Neighbors in fact are strongly inclined to cooperate, but they achieve cooperative outcomes not by bargaining from legally established entitlements, [...] but rather by developing and enforcing adaptive norms of neighborliness that trump formal legal entitlements. [...] [T]he end reached is exactly the one that Coase predicted: coordination to mutual advantage without supervision by the state."[21] In other words, people can use their local, context-dependent knowledge to establish new norms or informal rules that effectively address problems, and government oversight isn't necessarily required.

Additionally, Elinor Ostrom's work showed how communities and community organizations across the globe have established many different types of institutional arrangements for the creative management of natural resources, allowing societies to mitigate environmental problems. Ostrom realized that human-ecosystem interactions are complex and multifaceted, which led her to argue consistently against any one panacea for solving ecological problems.[22] As Ostrom argued in her Nobel lecture, "A core goal of public policy should be to facilitate the development of institutions that bring out the best in humans. We need to ask how diverse polycentric institutions help or hinder the innovativeness, learning, adapting, trustworthiness, levels of cooperation of participants, and the achievement of more effective, equitable, and sustainable outcomes at multiple scales."[23]

Much of Elinor Ostrom's work focused on small-scale issues, such as small fishing villages in Turkey, irrigation systems in Spanish villages, community-owned forests in Nepal, and alpine pastures in Switzerland. Some of Ostrom's critics have been skeptical of her findings because these bottom-up, communal governance institutions are not scalable to bigger domains, like an entire country. Instead, some critics have pushed for direct and comprehensive government interventions to address more complex cases of environmental degradation.

However, Ostrom's work shows that, in many cases, top-down government policies fail to achieve the stated goals, and they are often accompanied by negative unintended consequences. To address complex environmental problems effectively, we need a mixture of bottom-up and top-down approaches at a variety of scales and scopes. If Ostrom were asked how to solve any particular environmental problem, she would probably answer, "it depends" because of the different beliefs of people, their existing formal and informal institutions, and the exact nature of the problem.[24]

Legal scholar Robert Ellickson's work shows how sometimes there is a mismatch between what the formal laws say and the actual way in which governance operates on the ground. In fact, people may completely disregard the formal rules and stick mainly to unwritten, informal ones. As Ellickson has argued,

> The proposition that legal rules may lack bite is of particular importance to the legislators, lawyers, policy analysts, and others who aspire to be social engineers. These legal activists have been especially prone to exaggerate what the Leviathan can accomplish. For a wide variety of reasons, legal interventions can flop. To avoid the frustration of trying to influence what is beyond their reach, legal instrumentalists would be wise to deepen their understanding of the non-legal components of the system of social control. Indeed, one reason people are frequently willing to ignore law is that they often possess more expeditious means for achieving order. For example, neighbors in rural Shasta County [in northern California] are sufficiently close knit to generate and enforce informal norms to govern minor irritations such as cattle-trespass and boundary-fence disputes. This close-knittedness enables victims of social transgressions to discipline deviants by means of simple self-help measures such as negative gossip and mild physical reprisals. Under these circumstances, informal social controls are likely to supplant law.[25]

As we've seen so far, people act entrepreneurially in markets, government, and civil society. They are alert to and seize opportunities for change. With this in mind, the founders of FME coined the term "enviropreneur" as a portmanteau of "environmental entrepreneur." Although enviropreneurs have existed as long as entrepreneurship itself has existed, scholars are now studying how people can creatively find ways to overcome environmental problems through mutually beneficial exchange. As Anderson and Leal describe it,

> Environmental entrepreneurship is as much an art as it is a science. It takes different forms and does not lend itself to a formal structure, classification, or clearly defined rules. [...] The other distinguishing characteristic of enviropreneurs is their creativity, which allows them to discover new demands for environmental resources and to develop innovative ways to meet those demands. Enviropreneurs take risks and assume liabilities that others refuse.[26]

An enviropreneur isn't some special category of superhuman. Instead, literally anyone can be an enviropreneur, as long as they are alert to opportunities for improving environmental outcomes and seize those opportunities.

Enviropreneurs have an important role to play in the discovery and testing of new ideas. As discussed in Chapter 6, knowledge is dispersed and tacit, and the world is filled with countless "unknown unknowns." This is why enviropreneurs are so critical to addressing ecological issues. Different people have different types of knowledge and experiences, which allows them to be alert to and discover previously unrealized opportunities. The entrepreneurial market process can happen for environmental goods, just like it can happen for other common goods in the market, like food, houses, and electronics. As Anderson and Leal argue, "Just as individual species fill niches in ecosystems, entrepreneurs find market niches and specialize in production and marketing to fill niches. Successful entrepreneurship depends on the entrepreneur utilizing local knowledge and resources more efficiently than other individuals. As a result, inefficient resource use in markets and in ecosystems is crowded out in an evolutionary process where sustainability meets profitability."[27]

Since most environmental problems are, at their core, issues with property rights in some form, environmental entrepreneurs can find creative ways to work with property rights. They might find a better way to define property rights, or they might find a better way of enforcement. They might even create a whole new set of property rights that didn't exist before. For example, a conservation easement is just one creative way to use property rights for land preservation. A conservation easement is a voluntary legal agreement between a landowner and a government agency or land trust. Essentially, a landowner chooses to give up a part of their property rights to permanently limit future development of the land. The landowner can still own, use, and control their land. They can also sell it or pass it on to heirs. However, a certain part of those property rights is handed over, such as the ability to commercially develop the land or make major modifications. Participants in conservation easements often receive some financial benefit, like a tax break. In this way, conservation easements are a way to use private property rights to ensure conservation well into the future.[28]

The hallmark of an enviropreneur is seeking out win-win conservation efforts, instead of seeking to regulate behavior through government coercion.[29] It's important to emphasize that entrepreneurship in general, but especially enviropreneurship, can happen in profit or not-for-profit contexts.[30] For instance, the economists Virgil Storr, Stefanie Haeffele-Balch, and Laura Grube have done extensive research on commercial, social, ideological, and political entrepreneurs in post-disaster recovery situations, particularly after Hurricane Katrina in New Orleans and Superstorm Sandy in New York and New Jersey. These various kinds of entrepreneurs are all critical components of allowing communities to recover from natural disasters successfully. They argue,

> Entrepreneurs are a driving force behind the instances of social change. They are alert to opportunities to change society and are bold innovators who introduce new products, services, or ideas or establish new enterprises or revitalize existing ones to bring about social change. Entrepreneurs fill their role as agents of social change both in mundane times as well as in times of crisis.[31]

Their work on post-disaster recovery is directly relevant to environmental issues. Like post-disaster scenarios, environmental issues are time-sensitive and require many types of knowledge and feedback mechanisms to rapidly adjust to changing conditions. Entrepreneurs can use their context-specific knowledge to test out new ideas and revise those ideas as they receive feedback, thus helping to solve difficult problems. Entrepreneurs might even find ways of persuading people to change their customs and cultures so that they are more conscientious of environmental problems, like reducing their meat consumption, using plant-based meat substitutes, or switching to cultivated (i.e., cultured or lab-grown) meat. By shifting people's beliefs and attitudes, enviropreneurs serve as ideological entrepreneurs, which is the first stage of institutional change.[32]

Since it may be costly or difficult to solve environmental problems just with property rights and mutually beneficial exchange, some sort of public policies may be required. However, using government approaches to solve environmental problems is a double-edged sword. Taking the political route may cause resentment and conflict because people don't like policies that harm their wealth and livelihoods. Sometimes political ideologies take precedence over policies that yield effective results. Additionally, wielding political power creates a situation that is ripe for the issues in public choice economics, like rent-seeking and government granted privileges. The upshot is that when legal and political barriers are minimized, enviropreneurs have the freedom to innovate, discovering new approaches, technologies, or rules that nobody has yet thought of.[33]

Since entrepreneurs are such a critical component of solving socio-environmental problems, it is common sense to think that a larger population, leading to more entrepreneurs, would be better than a smaller population. However, several big debates raged in the twentieth century over whether the Earth was becoming too populated. One side of the debate saw humans as breeding too quickly, which would inevitably lead to environmental collapse, agricultural failure, and mass starvation. This line of thinking came from Thomas Robert Malthus, an English political economist who wrote in the late 1700s and early 1800s about the dangers of population growth in Great Britain. By the mid-1900s, a new set of thinkers called Neo-Malthusians extended Malthus's logic to the rapidly expanding populations across the globe. Two of the major figures in the Neo-Malthusian movement were ecologist and microbiologist Garrett Hardin and biologist Paul Ehrlich. The other side of the debate saw humans as ingenious, and they are "not only mouths to feed, but also hands to work and brains to think of solutions to pressing problems."[34] Two of the major thinkers on this side were economists Julian Simon and Elinor Ostrom.

Hardin is famous today for being the first person to articulate the logic of the tragedy of the commons, but besides that contribution, much of his work centered on natural resource management and population-control policy. Like many other scholars of his time, his views on population control were shaped by prior eugenicists and Malthusian thought. In Hardin's view, human population growth was the ultimate source of environmental degradation. As the scholars Pierre Desrochers and Joanna Szurmak describe, "In [Hardin's] view, any well-meaning policy—whether it involved research on alternative foodstuffs or poverty relief—that resulted in an increase in human numbers was ultimately self-defeating. Politically, Hardin was hostile to both laissez-faire free market policies and communism if the outcome was greater human encroachment on the natural ecosystems."[35]

However, Desrochers and Szurmak acknowledge that Hardin's rhetoric shifted over time as it became apparent that the dire consequences that he originally feared were unlikely to happen. However, for the rest of his career, Hardin "always maintained that population growth in the context of finite resources could only result in disastrous outcomes and, because of this, mandated a severe curbing of individual freedom."[36]

Perhaps the more famous Neo-Malthusian arguments came from Ehrlich, who was very pessimistic about population growth. Ehrlich and other scholars like him were experiencing the increasing environmental awareness that was taking place in the 1960s and 1970s. This awareness introduced new fears about resource exhaustion, the depletion of non-renewable energy sources, the increasing scarcity of metals for industrial uses, and the potential for mass starvation.[37] Ehrlich's book *The Population Bomb* was published in 1968, and it sparked international alarm over the potential of global famines due to human population growth. Ehrlich, and those who subscribed to his style of thinking, thought that populations had to be checked by voluntary or coercive means, otherwise humans would outgrow the limited supply of food and natural resources, which would result in irreparable environmental damage, yielding famine and societal collapse.[38]

Ehrlich and the other Neo-Malthusians argued that there were undeniable limits to economic growth and population growth because natural resources are obviously limited. All nonrenewable resources, like coal or iron, are finite, and when we run out, we run out. Similar to Ehrlich, environmental scientist Donella Meadows and her colleagues wrote *The Limits to Growth* in 1972, which claimed that the world's natural resources would be depleted in only a few decades, and the entire global population would be in danger. If economic growth were to continue on the same trajectory, said Meadows and her coauthors, more resources would be consumed through mass production, which would intensify pollution while also increasing the human population. A larger number of people would live in an increasingly polluted environment, accompanied by an evermore precarious agricultural system. The general quality of life and health of people would decline, leading to cataclysmic outcomes for humanity, sooner or later.[39]

Julian Simon was the main intellectual opponent of Ehrlich and the other Neo-Malthusians. Simon's scholarship argued that the ultimate resource "does not reside under the ground (natural resources), or even in the accumulated wisdom and knowledge in books and scientific journals (human capital), but in the imagination of people."[40] As Simon saw it, regular people observe the world around them and have context-dependent knowledge. Since people purposefully act to solve the problems they encounter, they can use their particular knowledge of time and place to create new technologies and rules that improve the human condition. In other words, Simon's core claim was simple: More brains lead to more ideas, which then leads to more innovations, which then improves both the human condition and environmental outcomes.[41] Humans can innovate themselves out of most predicaments. Simon wrote a book titled *The Ultimate Resource* to explain why he saw the human mind as our most precious commodity.[42]

Ehrlich and Simon were both true believers in their own positions, and they engaged in a famous public wager so that both of them had to put their money where their mouths were.[43]

In their wager, Simon and Ehrlich agreed that Ehrlich would choose a $1,000 basket of raw materials that he expected to become less abundant and more expensive. Ehrlich would also then choose a timeframe, and at the end of that timeframe, they would calculate the inflation-adjusted price of those materials. Ehrlich would win the wager if the real price of the "basket" of materials was higher at the end of the period, indicating the materials had become more precious. However, Simon would win the wager if the price was lower.[44]

In the basket of materials, Ehrlich chose copper, chromium, nickel, tin, and tungsten. The two scholars wagered $1,000 for a period of ten years, and they wrote up an official contract on October 6, 1980. A decade later, in October 1990, they looked at the real price of the basket of metals, adjusted for inflation, and Simon won the wager. When adjusted for inflation, the price of the basket of metals had fallen by 36 percent, which aligned with Simon's predictions and went against Ehrlich's. True to his word, Ehrlich wrote Simon a check for $576.07.[45]

In the years since the famous Simon–Ehrlich Bet, scholars have continued to debate and quibble about who was actually correct. If you change the timeframe or basket of materials in certain ways, Ehrlich would have won. However, changing the timeframe or the basket of materials in other ways would have made Simon win to a larger degree. Now more than thirty years have passed since the wager, and the evidence of a larger trend is relatively incontrovertible. Just as Simon had argued, resources are becoming more, not less, abundant, and what's more, the resources have become cheaper in terms of the amount of labor-time it takes to buy them.

Since the beginning of the twentieth century, the world has experienced many catastrophic wars, debilitating famines, and harsh economic downturns. But it has also experienced more technological improvement than at any other time in human history. In 1900, the world's entire population was just over 1.5 billion people. By 1970, the world had reached 3.6 billion people. In 2024, the world is just over eight billion people. More than five times as many people are alive as compared to 1900, but we are not on the verge of mass starvation. We have more food and resources than ever before, and the share of the world in abject poverty is the lowest share of the population that it has ever been. Simon was correct that resource constraints do not limit economic progress, but instead, more minds lead to more creative ways of resolving those constraints.[46] More brainpower gives us a greater likelihood of new beneficial inventions, and more brainpower allows us to see more ways to combine past technological developments in new ways. Thus, technology and brainpower are synergistic because advances are built on the combinations of old ones.[47]

The economic tools discussed in this book can help us understand why Simon was eventually proven correct and Ehrlich was proven wrong. Simon understood the role of the market process, how technological innovation comes about, the social function of entrepreneurship, and how knowledge is discovered, aggregated, and communicated. The problems posed by population growth and economic growth were opportunities for entrepreneurs and innovators to find solutions, just like they have found solutions to countless other problems over the centuries. Since people are imaginative and creative, Simon saw that they can push the frontier of what we thought was possible, opening up opportunities to meet the needs of a growing population. When materials and resources become scarcer, the prices rise, and with higher prices, producers have a strong incentive to increase the production of those materials and resources or find viable alternatives. Entrepreneurship is seeing how to increase production or discovering what viable alternatives are. Each time we try something new, a new environmental problem might arise, but that provides a new opportunity to provide people with the environmental benefits that they demand.[48]

Simon's work was important because it highlighted the real-world evidence that many of the dire predictions from many environmentalists were overblown or misguided. On the theoretical side, Simon tried to redirect their thinking away from environmental pessimism to a more rigorous study of the ingenuity of humanity.[49] As economists Peter Boettke and Christopher Coyne argue,

> Simon's faith in ordinary people was not a naïve faith. Instead, it was grounded in his observation that, even in the most difficult of circumstances, creative and clever human actors can use their imagination to adapt and adjust, to learn and innovate, and, in so doing, solve problems and spur progress. Progress is not the result of giving power to elites who purport to be extraordinary, but instead of empowering people to live their lives and pursue their bold conjectures about the world.[50]

In short, when institutional rules grant ordinary people a fairly broad degree of freedom, they can accomplish extraordinary things, as history has shown us time and again.

Humanity's ingenuity and entrepreneurial spirit can solve even the biggest of environmental problems. For example, economists Peter Jacobsen and Louis Rouanet have argued that climate change will lead entrepreneurs to find new ways to deal with higher temperatures that are not yet known or identified. They continue,

> More efficient air-conditioning will be created, entrepreneurs will innovate by building more energy efficient construction materials, technologies will be developed to make moving inland cheaper, seeds will be developed to enable crops to resist higher temperatures, etc. Population growth will give extra incentives for entrepreneurs to find new factors of production complementary to labor and thereby make it more productive. The fall in the existing supply of fossil fuel will give extra incentives to find substitutes to fossil fuels and fuel-economizing technologies.[51]

One related concept to Simon's way of thinking is the "Environmental Kuznets Curve" (EKC), which shows a general relationship between economic development and environmental impacts. The concept of a Kuznets Curve was named after Nobel Laureate Simon Kuznets, who found a relationship between the stage of economic development and the level of inequality in a society. As a country first starts to develop economically, levels of inequality rise, but at some point, the level of inequality peaks and then begins to decrease. The peak and the decrease in inequality arise because a larger proportion of people in a society benefit from industrial and technological advances, and they become more productive, making them wealthier. Additionally, governments may create social safety nets and education systems that reduce wealth gaps. If we were to think of the level of inequality graphically, the increase, peak, and decrease of inequality create a rainbow-shaped curve.

Using the same logic as the Kuznets Curve, other scholars have noticed the same pattern with ecological outcomes. When a country is relatively poor, its people don't consume many resources, and they have a relatively low environmental impact. Then, once a country starts to develop economically and becomes wealthier, its environmental degradation increases. After a while, the environmental degradation reaches a peak and then starts to decrease because a country's citizens and their government representatives are wealthy enough to place more weight on their care about the environment. Wealthier people have the luxury to invest in cleaner technologies and implement stronger environmental regulations, leading to reductions in pollution and environmental damage. During the earlier stages of development, many people are willing to trade off environmental concerns for economic concerns because the economic concerns are more pressing to them (see Figure 11.1).

The rainbow-like shape of the EKC is tied closely to Simon's arguments about "the ultimate resource." As a society grows economically, it can improve human welfare on a number of margins, like life expectancy, nutrition, income, and literacy, and it can also solve other important ecological problems, like air pollution, unsafe drinking water, and waste management.[52]

It's important to note that Simon's arguments are very dependent on institutions. Economic growth doesn't just happen in a vacuum. Economists have seen that certain patterns of formal and informal institutions, as well as cultural beliefs, lead to economic growth, and without that constellation of institutions, economic growth is unlikely.[53] Economist Vincent Geloso has

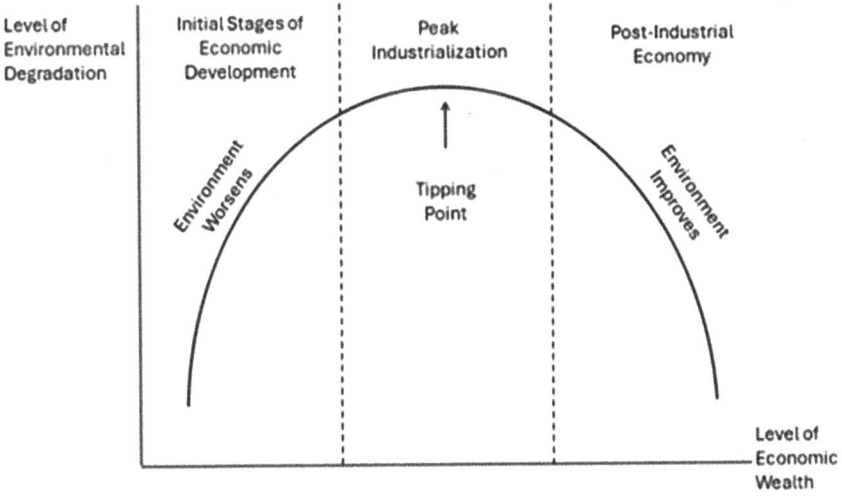

FIGURE 11.1 *Environmental Kuznets Curve.*

argued that Simon's insights about the ability of people to innovate their way out of large-scale problems is "always conditional on institutions. High-quality institutions hasten the move along the EKC and minimize the environmental damages produced before reaching the top of the EKC. Low-quality institutions do the opposite."[54]

At first glance, Simon's insights don't always seem to hold true. Looking at the environmental degradation in the Soviet bloc, Simon did not blame those problems on the people because they were inferior or lesser in some way. Instead, Simon's view was centered on the incentives created by a particular institutional framework. In fact, when institutions create perverse incentives, government actors can become the most destructive force on environmental quality. In the Soviet context, the drying and the dying of the Aral Sea were caused almost exclusively by the perverse incentives of the formal institutions. In other contexts, state-caused environmental destruction is still a major issue, and "bad" institutions create new environmental problems or amplify existing ones.[55] In fact, one of the most polluting industries in the world is national defense, which is driven entirely by governments.[56]

In sum, economic tools can illuminate ways in which people can discover creative and innovative solutions to environmental conflicts. Governments, markets, and civil society can use multiple approaches, each of which will have its own strengths and weaknesses, costs and benefits. Government legislation and regulations have long been the go-to for environmental concerns, but as public choice theory shows, government approaches are a double-edged sword because they can devolve into negative-sum games. To minimize the potential of that sword hurting ourselves, we should look for ways to use government policies at a variety of levels and locations so that we can experiment with a wide range of policy approaches. Many environmentalists have neglected the ability of the entrepreneurial market process to lead to discoveries of more effective and cost-efficient means of solving environmental problems. Another underappreciated means to address ecological issues is through civil society organizations, such as clubs and associations, which have a track record of finding workable, peaceful solutions. In short, there are many creative ways to use property rights, prices, and exchange to incentivize people to act in environmentally beneficial ways, leading to positive-sum outcomes that benefit everyone involved. We just have to allow people the flexibility to experiment with different systems.

12

Species Conservation

Species across the globe are endangered with extinction, and humans have caused many extinctions in the last few hundred years. Some scientists are claiming that the Earth is in the midst of its sixth "mass extinction," the previous one being the extinction event that killed off the dinosaurs. Current extinction rates are far above the "background" rates of a few hundred years ago. Scientists are still debating whether the current era technically fits the definition of a mass extinction, but regardless, many species of organisms are on the brink.[1]

Many people feel that it is a moral responsibility of humanity to keep species from dying off. Additionally, biodiversity has pragmatic benefits, like the discovery of new medications or food sources, and we lose those benefits if species are gone. For both moral and pragmatic reasons, people have engaged in a variety of different methods of species conservation, including government, markets, and civil society. This chapter looks at two examples: the conservation of the greater sage-grouse under the Endangered Species Act and the conservation of rhinos in many African countries. These examples show how successful or unsuccessful species conservation must focus on the different incentives that different institutional arrangements provide.

Governments have created laws and regulations to stop the extinction of species, and ideally, to allow those populations to flourish again. One of the most famous of these laws is the Endangered Species Act (ESA), which

President Richard Nixon signed into law in 1973. The law protects individual organisms of an officially designated species from being "taken"—a legal term that prohibits anyone from doing anything "to harass, harm, pursue, hunt, shoot, wound, kill, trap, capture, or collect, or to attempt to engage in any such conduct." The ESA in the United States has been the source of much conflict since it became law in 1973, especially since listing a species as "endangered" or "threatened" under the ESA significantly impacts how people can use private and public land.

The law has been amended a few times, particularly in 1982, when Congress required federal agencies to designate a critical habitat when an endangered or threatened species is listed. A critical habitat is a geographical area that contains "the physical or biological features that are essential to the conservation of endangered and threatened species and that may need special management or protection."[2] Critical habitat designations affect only federal agency actions or federally funded or permitted activities. Although critical habitat designations don't directly affect activities by private landowners, it is not always so clear cut. For many people who use or lease federal land, such as millions of ranchers and farmers in the western United States, the critical habitat designation directly affects their livelihoods.

The ESA can limit or restrict how people use their private land or leased public land, but there is no direct compensation mechanism in the law. This has led to a few different negative unintended consequences, including preemptive habitat destruction and a phenomenon called "shoot, shovel, and shut up."

With preemptive habitat destruction, economists Dean Lueck and Jeffrey Michael explain that "landowners with potential endangered species habitat may have the incentive to preempt the ESA by destroying those characteristics of the land that would attract the species. Such preemptive activity would be a completely legal land-use decision spurred by the potential for costly regulations."[3]

In other cases, landowners or lessees of public lands fear the regulatory consequences of the ESA, which can lead them to preemptively harm endangered species to avoid the restrictions. In the most extreme cases, these landowners or lessees act in their rational self-interest to "shoot, shovel, and shut up," which removes a potential detriment to their livelihoods.[4] Ironically, the very law that is meant to protect endangered species ends up incentivizing landowners and lessees to harm them instead, at least in some cases.

However, despite the lack of a direct compensation mechanism in the ESA, other government policies and actions from civic associations and nonprofits have mitigated the negative unintended consequences. Some states offer tax incentives or credits to landowners who engage in conservation activities that protect endangered species and their habitats. For many years, landowners have used Safe Harbor Agreements, which are voluntary agreements between landowners and the US Fish and Wildlife Service (FWS) that encourage landowners to manage their land in ways that benefit endangered species. In return, landowners receive assurances that no additional regulatory restrictions will be imposed on them if their conservation efforts result in an increase in the population of endangered species on their property. Essentially, these agreements can protect landowners from future liabilities under the ESA. As of 2024, the FWS combined Safe Harbor Agreements with a similar approach called Candidate Conservation Agreements with Assurances, and the new agreement is called a Conservation Benefit Agreement.[5]

Habitat Conservation Plans are another important part of the ESA. The FWS can negotiate with landowners to develop plans that specify how landowners will minimize and mitigate the impact of their activities on endangered species. In exchange, landowners can obtain an "incidental take permit" from the FWS, which essentially authorizes landowners to harm a limited number of individuals of an endangered species or their habitat. This may not be a direct form of compensation, but it does give some flexibility within a relatively strict and invasive law.[6]

Land acquisition programs, like Recovery Land Acquisition Grants, allow the federal government to purchase land or conservation easements from private landowners, which can then be used to protect critical habitats for endangered species.[7] Thus, landowners are compensated for the value of the land or the easement, making them less likely to harm species or their habitats. In other words, these various programs, including Conservation Benefit Agreements, Habitat Conservation Plans, and Recovery Land Acquisition Grants help shift the incentives that people face. Without these programs, the strict nature of the ESA can turn the presence of an endangered species into a liability, but these alternative policy arrangements reduce the costliness of having an endangered species around.

Scholars debate whether the ESA should be counted as a failure or a success. A relatively small percentage of all species that have been listed as "endangered" under the law have gone extinct, so in that sense, it is a success. However, relatively few species have had their populations sufficiently boosted to be removed as an officially listed endangered species, which would make the law a failure. With the ESA's results existing in a Limbo between success and failure, other scholars have called for revised and innovative approaches to species conservation.[8]

One of the most instructive episodes for thinking about how to revise and innovate within the ESA was the massive conservation effort of the greater sage-grouse in the western United States. The greater sage-grouse is a ground-dwelling bird that lives in the arid "Sagebrush Sea" of eleven western states. The species is a favorite of bird watchers because of its unique appearance and its elaborate mating rituals. The males attempt to attract a mate by displaying their spiky tail feathers like a fan while also inflating the two yellow air sacs on their chests. These air sacs create a loud popping noise, which enhances their mating dance.

Although the sage-grouse is a favorite among bird watchers, its habitat is among some of the most valued land in the West. Human development, including urban expansion, large-scale ranches, and resource extraction, is happening rapidly in the Sagebrush Sea. In fact, some of the fastest growing states are home to the greater-sage grouse, including Utah, Idaho, and Nevada, which were three of the top-five fastest-growing states from 2010 to 2020.[9]

Due to a declining population and a shrinking habitat, the FWS declared in 2010 that the greater sage-grouse warranted protection as an endangered species under the ESA. In that announcement, however, the FWS also announced that it would not yet formally list the species as "endangered" or "threatened" due to a lack of resources. The FWS's determination was that an official designation under the ESA was "warranted but precluded." In other words, FWS officials indicated that they wanted to list the sage-grouse under the law, but they decided to wait to see if the populations could be increased before moving forward with an official action.

Federal, state, and local policymakers, as well as landowners and some nonprofit groups, were afraid of the species gaining the full protection of the ESA. The sage-grouse's habitat extends across tens of millions of acres of private and public land, which would disrupt many facets of economic development and recreation. Some feared a worst-case scenario, meaning that the strictest protections would lead to huge economic repercussions of tens of billions of dollars.

Some conservation groups, however, lauded the potential listing of the greater sage-grouse because the bird is an "umbrella species." The concept of an "umbrella species" refers to the conservation of one species indirectly leading to protection of many others. The sage-grouse lives nearly everywhere in the Sagebrush Sea, including the Great Basin, the Colorado Plateau, and the mountain valleys of the Rockies. If the habitat of the sage-grouse were more strictly protected by the ESA, then many other forms of wildlife would benefit under its "umbrella," including pronghorn, mule deer, elk (aka, wapiti), pygmy rabbits, and many other bird species. Therefore, the sage-grouse and its habitat mattered more than just themselves—the species was a convenient way to conserve other ones.[10]

The number of people in the western United States that feared the listing of the bird—especially state and local policymakers—far outnumbered the ones who supported it. Due to pushback from these political leaders, FWS officials allowed and encouraged a wide variety of conservation approaches to avoid a full listing under the ESA. Between 2010 and 2015, policymakers at the federal, state, and local levels, plus market firms, private landowners, nonprofit organizations, academics, and conservation practitioners, implemented creative and innovative ways of conserving the sage-grouse. In short, these various entities experimented with a highly polycentric approach to species conservation. Ultimately, in 2015, FWS officials chose not to list the greater sage-grouse because they determined that the species' population was sufficiently large and stable.[11]

During that time period, officials at the Bureau of Land Management (BLM) and the US Forest Service (USFS) drafted new management plans and adopted them in 2015, which expanded coordination between the two agencies. These new plans also provided technical assistance and financial support for conservation on private lands. When creating these new plans, officials in the BLM and the USFS received input from a wide range of stakeholders, including farmers, ranchers, energy developers, state fish-and-wildlife agencies, and many others.

Additionally, the BLM and the USFS cooperated with the Sage Grouse Initiative—part of the Natural Resources Conservation Service in the US Department of Agriculture—which was intended to restore roughly 4.4 million acres of sage-grouse habitat while also allowing economic development on federal lands. The Initiative focuses on "win-win solutions to target voluntary, incentive-based conservation that improves agricultural productivity and wildlife habitat on working lands."[12] During this time period, more than 1,100 private individuals voluntarily participated in incentive programs and community support, leading to increased protections for sage-grouse habitat and boosting sage-grouse populations. Through the Initiative, ranchers secured conservation easements, promoted deep-rooted perennial grasses to keep the range weed-free, removed conifers that threaten sage-grouse habitats, performed wetland restoration projects, and increased visibility on fences to reduce deadly collisions with sage-grouse.[13]

Political leaders in western states, especially governors, spearheaded a variety of new programs and initiatives to boost sage-grouse numbers and protect critical habitats. One of the most successful approaches was in Utah due to the sage-grouse conservation plan's highly polycentric structure. Many governance decisions were devolved to very local levels and included direct input from private landowners, commercial associations, and nonprofits. Since a large portion of Utah's greater sage-grouse population resides on private lands—some estimates place it around half of the population—the buy-in from private landowners and their commercial associations was a critical component of making the conservation plan successful. Action from private landowners and their commercial associations was especially important because they had knowledge about their unique circumstances and about the local habitat. By decentralizing control to some degree, they could leverage their knowledge effectively for conservation. Additionally, since they were involved in the decision-making processes, they saw the newly created rules as legitimate, which made them more likely to actively participate.

Utah's sage-grouse conservation plan was a complex undertaking with many different facets, but one part of the plan that deserves to be highlighted is the "local working groups." These groups are composed of relevant stakeholders within a relatively small geographic region. The stakeholders included federal, state, and county officials, plus local landowners, academics, commercial associations, nonprofits, conservation groups, and other interested parties. Utah State University's extension program facilitated these local working groups through its Community-Based Conservation Program, and university-affiliated researchers guided the groups with their expertise while also listening to the feedback from the other stakeholders.

The system of local working groups was first created in 1996, which predated the FWS's 2010 decision of "warranted but precluded," but it was expanded and became even more important after 2010. Within each of the sub-regions in Utah, the groups drafted their own conservation plans, which were tailored specifically for the local environmental conditions and the local communities' preferences. Utah still uses local working groups today, and there are ten across the state. Several other western states have adopted a similar approach to Utah, allowing lower levels of governance and bottom-up solutions. More than sixty local working groups exist in the West.[14]

The future of the greater sage-grouse is uncertain because of the continuing human population growth in the western states, as well as ranching, resource extraction, and climate change. However, the episode between 2010 and 2015 shows how a truly polycentric governance approach can align people's knowledge and incentives to engage in successful conservation. The success is rooted in the combination of new federal land-management plans, the win-win approaches of the Sage Grouse Initiative, various state policies, and collaborative local working groups. Rather than top-down directives and strict command-and-control approaches, species conservation can emerge from discussion, deliberation, and cooperation. Mutually beneficial arrangements were a key component of the story, including incentive programs, community support, and multi-stakeholder involvement.

The story of the greater sage-grouse is a lesson in intelligent policymaking through a polycentric approach. However, one of the most underappreciated means of saving species from extinction is through private property rights and markets. Scholars have been researching how some of the most beloved species have been brought back from the brink by using the power of rational self-interest, incentives, and mutually beneficial exchange.

Rhinos, as one of those beloved species, are one of the most iconic animals in the world. They are a prime example of "charismatic megafauna"—large animals that have a widespread popular appeal, such as elephants, pandas, lions, tigers, and eagles, among many others. In fact, these animals are so charismatic that people go to great lengths to get them, such as hunting them as trophies or using their body parts in traditional medicines.

Across Africa, rhinos of various species and subspecies were hunted to near extinction in the nineteenth century. The remaining populations were concentrated in a few national parks scattered across the continent. Governments struggled to keep poaching under control, and the populations continued to decline through the twentieth century in most places.

In Africa today, rhino ownership takes three general forms: private, state, or community-based landholders. Rhino ownership and landownership are not necessarily the same, but there can be overlap due to various institutional forms of property rights. For example, in some places in Africa, communities reside on state lands, and they have rights over the wildlife on that land. Depending on the specific institutional structures, landholders—whether private or communal—may be rhino custodians, but they may not necessarily own rhinos (or land, for that matter). When private owners and communal landholders have power to oversee rhinos, they can gain financially from wildlife-based tourism, regulated trophy hunting, wildlife sales, and meat sales.[15]

Thus, the big question is this: What are the most effective means for achieving a certain end? If the goal is to boost the population of rhinos and keep them from extinction, then should we favor the property-rights approach or the government-control approach? The evidence has shown that (for the most part) the property-rights approach has been more successful.

In the twentieth century, one of the largest rhino populations was in South Africa, but before 1991, all wildlife in South Africa was treated by law as unowned property. Wildlife was essentially under the control of the government, but individuals could not own these animals. The government of South Africa would sometimes sell off rhinos to private game ranchers who would then sell the opportunity to hunt the rhino as a trophy.

Using the lens of economics, we can analyze why the old rules of the government selling off rhinos one-by-one to game ranchers did not create incentives to protect or propagate the species.[16] Under the law, even if a game rancher paid for a rhino, the rancher could not claim compensation if the rhino left his property or was killed by a poacher. For example, in 1982, the Natal Parks Board—one of the government entities in South Africa—listed the price for a live white rhino at 1,000 South African rands. But the market price for a trophy rhino was selling for 6,000 rands, so any private landowner who bought a rhino from the Natal Parks Board had a very strong incentive to sell the rhino as quickly as possible to pocket a sixfold profit.[17]

In 1985, a private rancher in South Africa chose to auction off a few rhinos that it had received from the Natal Parks Board, which then prompted the Board to do the same. The next year, the Board auctioned six rhinos at 10,000 rands each. The Board then fully embraced the auction system, showing the market price for rhinos. By 1989, the price for a rhino peaked at 92,000 rands.[18]

Policymakers in South Africa were persuaded in 1991 to allow for the private ownership of any wild animal that could be identified, such as by a brand or an ear tag. The Theft of Game Act, which allowed for this private ownership, had the effect of creating a market for rhinos and also creating stronger property rights. With the combination of private property rights and market prices, the incentives for private ranchers fundamentally changed. It was in the self-interest of owners to breed rhinos rather than to shoot them quickly.[19]

In other words, when people own animals, they have a stronger incentive to breed the animals and protect them from harm. Think of this parallel example. Everyday, across the world, millions of cows, pigs, and chickens are slaughtered, but these species are not in danger of extinction. Why? The owners have a strong incentive to breed animals to keep the populations high so that they can continue their livelihoods in the long term.

The outcome of South Africa's new system was miraculous. In 1900, the southern white rhino was the most endangered of the world's five rhino species. Less than twenty southern white rhinos remained in a single reserve in South Africa. However, by 2010, southern white rhinos numbered more than 20,000, making it the most common rhino species on Earth.[20]

Today, South Africa has the largest number of rhinos in all of Africa, with roughly 81 percent of the white rhinos and 33 percent of the black rhinos, but there are some troubling trends. For most of its history, the largest populations of South Africa's rhinos were in Kruger National Park, but in the past decade or so, the park has become a poaching hotspot. Between 2011 and 2021, the white rhino population declined by 76 percent and the black rhino population declined by 68 percent in Kruger.[21] Across South Africa, the other national and provincial parks containing rhinos have reported declining rhino numbers over the past decade due to poaching.[22]

The story of the southern white rhino shows the importance of institutions and the incentives they create. Namibia has followed South Africa's lead by allowing the private ownership of wildlife, including rhinos. These two countries alone contain roughly three-quarters of the world's black rhino population and over 90 percent of the white rhino population. Kenya, Tanzania, and Zambia, however, took a very different approach. The black rhino, which is predominant in these three countries, cannot be owned privately. In 1960, about 100,000 black rhinos roamed across Africa, but by the 1990s, poachers had reduced that number to less than 2,500. The black rhino has had a population increase in many places across Africa since the 1990s, but this has required massive government investments, including armed guards that follow rhinos around to protect them from poachers.[23] The government-operated protection zones in Kenya and Tanzania still attempt to stop poachers with armed guards, but even with all those resources, the property rights approach in South Africa and Namibia has been more successful at boosting the populations.[24] This is because the property rights approach aligns the incentives of the owners with the propagation of the species.

However, one major complication with owning wildlife, especially if that ownership leads to trophy hunting, is that people disagree on their normative conceptions of right and wrong. Some people see trophy hunting as a moral evil. Even if trophy hunting is evil, that does not negate the existence of tradeoffs. The tradeoff of forbidding the private ownership of wildlife is that we are forbidding a potentially more effective means of propagating a species and keeping it from extinction. The field of economics can't tell us which tradeoff is right or wrong, but it can tell us that any choice we make will have an opportunity cost. The trick is coming to some sort of consensus on whether that cost is worth it.

Despite the declining number of white rhinos in national and provincial parks in South Africa, the estimated number of white rhinos on private land has steadily increased. In 2021, white rhinos on private land constituted 53 percent of the total white rhino population in South Africa, up from only 25 percent in 2010. Now, private landholders support the largest number of white rhinos, not only in South Africa, but on the entire continent. As for black rhinos in South Africa, roughly 25 percent are conserved on privately held lands.[25] In interviews conducted by scholars between 2016 and 2018, private owners and managers of rhinos said that rhinos are generally easy animals to keep, with only minor concerns regarding drought and the potential for inbreeding. The private owners and managers expressed that one of their main motivations is a feeling of contributing to rhino conservation in South Africa.[26]

To raise rhinos, prospective rhino entrepreneurs usually need to own a substantial amount of land. Landholders will make a rational calculation about the benefits of raising rhinos compared to their relevant land-use alternatives, such as agriculture. The most important part is whether public policies allow landholders to engage in a full range of decisions, rather than restricting them. Since legal institutions in South Africa allow private landowners to have full ownership rights to the wildlife on their properties, they have broad scope to engage in various forms of environmental entrepreneurship. However, in Kenya, people have more limited rights for consumptive wildlife uses, so revenues come largely from wildlife-based tourism, rather than trophy hunting, meat sales, or similar sales.[27] These restrictions limit the scope of potential environmental entrepreneurship that people can engage in.

The private ownership of rhinos, as well as other large game, is not an easy task, however. It may not be viable for all people who may want to engage in it. Large game usually requires a large land base, extensive intervening in the social lives of these animals (such as facilitating breeding), and sufficient resources (such as supplementary food, water, and shade). Additionally, since rhinos and other large game are mobile, adequate fencing is required, which is often an expensive capital investment. Thus, African game ranches have a large continuum of potential costs and benefits, making a viable private ranch dependent on many factors. Additionally, private ranchers may have to weigh the tradeoffs of managing a wildlife ranch mainly for conservation purposes or for maximizing financial profit.[28] Many of the ranchers who raise rhinos talk about their "double bottom line" of both earning a livelihood and preserving a species that they love.

In surveys conducted from February to April 2017, rhino owners admitted that they either frequently (54.6%) or occasionally (24.2%) considered removing rhinos from their lands due to the rising expenses and risks. A few of the rhino owners said that they rarely (9.1%) or never (12.1%) considered removing their rhinos. When asked about their motivations to keep rhinos, owners had a variety of reasons: A passion for rhino conservation (94.0%); rhinos attract visitors (57.6%); owners want to have as many native species as possible on their lands (45.5%); and rhino horn could become a potentially valuable investment (45.5%).[29]

In South Africa, the white rhino population on private land has experienced lower poaching rates than populations on government-owned land. It is not guaranteed that this trend will continue, but it seems likely that it will, based on the personal incentives of private landowners who directly benefit from the protection of their land and the animals on it. However, with poachers becoming more sophisticated over time and rhino populations shrinking on public land, poaching may be displaced to private lands.[30]

In recent years, private landholders in South Africa who own white rhinos have faced a number of hurdles. Some are divesting from white rhinos because the heightened threat of poaching has raised security costs, no longer making the raising of rhinos profitable. The COVID-19 pandemic, perhaps unsurprisingly, cut sharply into the numbers of tourists and trophy hunters, which made some operations unprofitable. However, there are some positive trends, and most rhino owners in South Africa have maintained stable populations. A recent survey found that roughly 15 percent of South African private rhino owners are actually investing in more rhinos, as of 2020.[31]

But, due to the rising security costs from poachers and the disruption of the COVID-19 pandemic, rhino numbers are still growing on private land, but those rhino populations are concentrated on a fewer number of larger properties that could cope with these shocks. If costs continue to rise or other shocks occur, it's likely that more private owners will disinvest from rhinos. Perhaps community-owned land and communities can be a helpful source to protect rhinos in the future, but costs will be the biggest factor. If funding, through donors or other revenue streams, is secured, the combination of government, private, and community efforts could continue to boost the overall rhino populations, and make a broadly inclusive conservation system.[32]

Poachers are driven to engage in this illegal activity because they make a rational calculation between the expected costs and expected benefits, dependent on their own personal tolerance for risk. As the international black-market price of rhino horn increases, the incentive to engage in poaching becomes stronger. As the price decreases, fewer people engage in poaching because the costs outweigh the benefits. In recent years, governments and private owners have tried ways to slow poaching through enhanced law enforcement techniques and dehorning rhinos, but these approaches haven't been successful in many cases—as the recent surge in rhino poaching in South Africa suggests. Some scholars are saying that more of the same law enforcement will be in vain, so policymakers, rhino owners, and conservationists will need to be more adaptive and creative.[33]

One of the most controversial discussions for rhino conservation is legalizing the rhino horn trade, which is currently banned by the Convention on Illegal Trade in Endangered Species of Wild Fauna and Flora (CITES). Rhino advocates have called the CITES agreement a "blunt instrument" of a top-down approach. However, the legalized but restricted trade of rhino horns may be possible under certain conditions. If the institutions are crafted carefully, private ownership of rhinos could better meet the demand of rhino horn with the supply in a sustainable way.[34]

Researchers have found that the majority of South African rhino owners support the legalized trade of rhino horn, assuming property rights are protected and poaching can be kept at bay. In a recent survey of rhino owners in South Africa, 81.9 percent of respondents strongly agreed that legalizing the global rhino horn trade would benefit rhino owners, and 81.3 percent said it would benefit rhino conservation. Relatedly, 63.2 percent thought legalizing the trade would reduce rhino poaching.[35]

Rhino owners and managers interested in legalizing the horn trade fear for the future of rhinos because of the rising costs of protecting them from poaching, which may cause them to disinvest in rhinos at some point. Owners and managers also mentioned safety concerns as a top priority.[36] By better understanding the financial and non-financial costs of rhino owners—including the risks that poachers pose to owners and their family—we can better understand whether legalization of the rhino-horn trade might be feasible.[37]

Of course, such a controversial move to legalize the rhino-horn trade will require more academic research and also buy-in from the international community. Amending the CITES conditions could have positive effects, but it could also have unforeseen consequences on local livelihoods and wildlife populations because of the complexity and interrelatedness of legal markets, illegal markets, ecosystem dynamics, and so on.[38]

Rhino owners and managers have consistently expressed their lack of trust and faith in the government to quell poaching and boost population numbers.[39] In a survey of rhino owners, a majority of respondents thought that the South African Department of Environmental Affairs either definitely (51.8%) or probably (25.9%) increases the risk of poaching events. The lack of faith in provincial environmental departments was similar, with owners saying those departments definitely (46.1%) or probably (27.5%) increase the risk of poaching.[40]

In essence, what is required is more entrepreneurial, innovative approaches to limit poaching and boost rhino populations. Perhaps part of the solution will be improving the supply, such as breeding rhinos specifically for their horns, which will lower the price on the market, thus reducing the incentive to poach. The other side is treating the source of the issue, which is the demand for horn. This would be a role for social and cultural entrepreneurs, who may be able to influence people's preferences so that they don't want rhino horn. By shifting demand, market prices for rhino will also decrease, leading to less poaching.

In terms of government action for rhino conservation, the concept of opportunity cost is important. If policymakers want to be truly effective at rhino conservation, their approaches need to be flexible and adaptive because poachers are becoming more advanced. By doubling down on traditional techniques of enforcement, government officials direct resources away from other potentially more advantageous conservation activities. The opportunity cost of missing out on more effective conservation activities could be huge.[41]

While private wildlife ranches in Africa have been successful at boosting rhino numbers, other countries have created policies for community-based natural resource management. In South Africa, Namibia, and Zimbabwe, policies have allowed communities to benefit from the management of rhinos, allowing the revenues from wildlife hunting and tourism to flow to communities as a whole. South Africa's community-based conservation is less well developed than Namibia and Zimbabwe because of its unique history of concentrated private land ownership due to the Apartheid regime. South Africa's National Biodiversity Economy Strategy has been allowing indigenous communities to have a larger say over the management and revenues from rhinos, remedying at least some of the many injustices of Apartheid.[42]

Nonprofits and community initiatives also play an important role in rhino conservation. For example, in Namibia, Wildlife Credits is an initiative that crowdfunds donations to support payments to communal conservancies linked to conservation performance, including rhino sightings and monitoring. Another example is One Africa, which uses crowdfunding to allow members of the public to become shareholders who support conservation businesses, including rhino conservation.[43]

Overall, species conservation—whether for sage-grouse, rhinos, or anything else—will require action on the part of governments, markets, and civil society. Each of these areas has distinct advantages and disadvantages, so the goal should be to use the advantages to the greatest extent possible, while also minimizing the disadvantages. For rhinos, private ownership and markets have been hugely successful in boosting population numbers, even in the presence of more cunning poachers. However, not all species are as "charismatic" as rhinos, so private ownership and markets are not a panacea for conservation. Government policies and regulations may be necessary for conservation, but we must remain mindful of the unavoidable tradeoffs and potential unintended consequences. Civil society groups can play important roles in supplementing the efforts of both markets and government actions, as well as providing checks and balances when those actions might be harmful. There is not one correct way to approach species conservation, which means that we need processes of trial-and-error to discover the right combination of institutions that provide effective results.

13

Land Conservation

A long-standing assumption is that large-scale land conservation or preservation requires government action, like national park designations. Yellowstone was set aside as the world's first national park, or as the law that created the park says, as a "pleasuring-ground for the benefit and enjoyment of the people." Millions of people now visit Yellowstone every year, crowding to see its famous geysers and hot springs, as well as its abundant wildlife, pristine lakes, soaring mountains, open meadows, deep canyons, and thundering waterfalls. Starting with Yellowstone, governments across the globe have designated hundreds of national parks, with millions upon millions of visitors. In the United States alone, the National Park Service counted 325.5 million recreational visits to the National Park System in 2023.[1]

Besides national parks, there are many other types of conservation designations for public land, including wildlife refuges, official wilderness areas, and national monuments. For example, the US Fish and Wildlife Service oversees roughly 590 wildlife refuges across the country, spanning from Alaska's 19.6-million-acre Arctic National Wildlife Refuge to Minnesota's 0.57-acre Mille Lacs National Wildlife Refuge.[2] In 1964, Congress passed the Wilderness Act, which created the National Wilderness Preservation System. In federally designated wilderness areas, no one is allowed to use any motorized equipment, build any structures, or cut any trees. The Wilderness Act poetically defines a wilderness in this way: "A wilderness, in contrast with those areas where man and his own works dominate the landscape, is hereby recognized as an area where the earth and its community of life are untrammeled by man, where man himself is a visitor who does not remain." Many federally designated wilderness areas are large and remote, like central Idaho's 3.9-million-acre complex of the Frank Church–River of No Return Wilderness, Selway–Bitterroot Wilderness, and Gospel Hump Wilderness. Other wilderness areas, however, are relatively small and border large metropolitan areas, like Utah's Mount Olympus Wilderness, Twin Peaks Wilderness, and Lone Peak Wilderness, which are immediately adjacent to the suburbs of Salt Lake City and its famous ski resorts.

The global popularity of national parks leads to some difficult tradeoffs with conservation, however. Each year, hundreds of millions of people visit national parks, posing a tension between ecological conservation and public recreation. Government officials across many countries have adopted best practices to ensure that recreation and conservation are as compatible as possible, but as more people come to national parks, it becomes more challenging to ensure the preservation of wildlife and landscapes.[3] As one park official said, "The last thing you want to do is love your park to death!"[4] With millions of annual visitors in a national park, long-lasting damage may be unavoidable.

Policymakers have the difficult task of deciding how to balance the tradeoffs. In some national parks, policymakers have instituted daily quotas and ticketing systems to push the scales in favor of conservation over recreation. Many American national parks required timed entry tickets for admittance to popular sites. For example, during peak summer travel time, Colorado's Rocky Mountain National Park, California's Yosemite National Park, and Utah's Arches National Park require timed tickets for entry. Popular hiking trails or scenic drives in some national parks also require a reservation, such as Old Rag Mountain Trail in Virginia's Shenandoah National Park, Angels Landing Trail in Utah's Zion National Park, or Going-to-the-Sun Road in Montana's Glacier National Park. Without any sort of restriction, the millions of yearly visitors to these places could end up harming the very sites that a national-park status is meant to protect.

Sometimes, well-intentioned visitors to national parks even end up harming the wildlife they came to enjoy. In 2016, National Park Service officials had to euthanize a baby bison in Yellowstone after a man placed it in the back of his vehicle because he thought the calf was too cold.[5] In 2023, another visitor to Yellowstone attempted to help a bison calf that had been separated from its mother. Park rangers tried several times to reunite the calf with the herd, but the calf lost its fear of cars and people, creating a hazard. Ultimately, park staff had to euthanize this calf too.[6]

Keeping people from loving parks to death is a difficult task, but the policymakers who oversee that task also rationally respond to their incentives. For example, in Yellowstone between 1872 and the 1960s, managers in the National Park Service actively managed the number of elk within the park's boundaries. They engaged in active culling and the translocation of large numbers of elk from the park to other areas in the western United States. Too many elk means that ecosystems would be damaged by overgrazing, so park officials saw this as a necessity to fulfill their mandate of conserving the park for future generations.

However, in the 1960s during the early stages of the environmental movement, mass media reported on increased elk reductions that the National Park Service had undertaken in Yellowstone. This news sparked a public outcry against active elk-population management, and members of Congress were hearing from many constituents. Political pressure from Congress pushed park managers to make substantial changes in management policy, which limited the active management of the elk population, causing it to grow.

There were cascading effects through the Yellowstone ecosystem. The artificially large elk populations damaged the willows and aspens throughout the park, which affected other species. For example, the lack of willows caused the beaver population to decline, which meant fewer beaver ponds. Moose are dependent on beaver ponds, so the moose population also declined. Many of the issues with elk-caused overgrazing weren't resolved until the National Park Service and US Fish and Wildlife Service reintroduced wolves to Yellowstone in 1995, which used natural predation to control the elk numbers.[7]

In addition to national parks, the management of officially designated wilderness areas is more complex than it may seem at first glance. In the US context, officials in federal land agencies have struggled to balance the difficult tradeoffs associated with stewarding wilderness areas because the provisions of the Wilderness Act make it nearly impossible to engage in active management. At first, it may seem like an official wilderness area would be a logical place to engage in the conservation or restoration of endangered species. However, designated wilderness areas cannot be modified, and mechanized equipment is not allowed to be used within the boundaries.

For example, in 2007, the US Fish and Wildlife Service authorized the construction of two water installations in Arizona's Kofa Wildlife Refuge and Wilderness Area to boost the declining bighorn sheep population. When Wilderness Watch and several other conservation groups learned of the construction of water tanks within Kofa's designated wilderness area, they filed a lawsuit challenging the activities, on the grounds that it violated the Wilderness Act and National Environmental Policy Act. The Wilderness Act forbids the construction of permanent structures, which would include the water tanks. It also prohibits the use of motorized vehicles inside wilderness boundaries, so the backhoes and trucks used to transport and install the tanks would be a violation of the law. In 2008, the US District Court for the District of Arizona upheld the construction of the water tanks, but in 2010, the Ninth Circuit Court of Appeals reversed the district court's decision, ruling that the water tanks did not meet the minimum necessary requirements for administering the area as wilderness.[8] In short, the battle over the water tanks in the Kofa Wilderness shows that federal land managers must justify any actions that might go against the Wilderness Act, even when those actions are meant to conserve land and species. The constraints of the Wilderness Act might sometimes have the unintended consequence of limiting our ability to engage in conservation measures that many environmentalists might value.

Even though governments can be and often have been successful in preserving land, government action isn't the only means. Private action, through markets and civil society, can be, and has been, successful. Although there are many possible examples to examine, perhaps one of the most interesting and largest ones is the American Prairie Reserve (APR), which is both a private, nonprofit organization, as well as a location. The organization's core goal is to restore hundreds of thousands of acres on the Great Plains of north-central Montana to its condition in the early nineteenth century. This area is where the Lewis and Clark Expedition passed through over two hundred years ago, and their journals documented that the shortgrass prairies were teeming with wildlife.

When the full project is completed, APR's land will contain large herds of bison and other wildlife, such as deer, elk, pronghorn antelope, and even bears and wolves. It will essentially be a privately owned national park, and public visitors can access the land for free and engage in various types of recreation, including hiking, driving, photography, hunting, camping, and biking. Alison Fox, the CEO of APR, said that "the overall goal is about 5,000 square miles, or 3.2 million acres of intact grasslands, comparable to the size of the state of Connecticut and also comparable to Glacier and Yellowstone National Parks combined."[9] Fox said, "Across the West, you see more and more 'no trespassing' signs. We've talked a lot about wildlife and wildlife habitat, but certainly providing access for people to appreciate, learn from, and recreate on this land is a really important part of what we're doing as well."[10]

APR is using an innovative way of conserving this large-scale landscape. Its core tenet is to engage in willing-buyer, willing-seller transactions to stitch together a large landscape that can support large numbers of wildlife. APR raises funds from private donors who support their cause. APR then approaches ranchers in north-central Montana and asks if they would like to sell their privately owned land. If landowners do not want to sell, they have no obligation to do it. In this way, APR is a clear example of using mutually beneficial exchange to accomplish environmental goals.[11] APR has purchased thirty-four ranches from the $200 million it has raised from more than 4000 donors, including Wall Street financiers and technology billionaires. APR's CEO expects that the entire project will take hundreds of millions of more dollars and decades of time to complete the full vision.[12]

The private land is intermixed in a patchwork of federal land owned by the Bureau of Land Management (BLM) or state lands managed by the Montana Department of Natural Resources and Conservation. Public grazing land is leased long-term to ranchers who own adjacent private land. Under the existing federal and state rules, the new owners of private land, including APR, can acquire the same grazing leases, if the new owners uphold applicable rules and regulations.

Sean Gerrity, the founder of APR, described the project in the following way:

> I grew up in Great Falls, Montana. My parents used to bring me out, and we would camp right on the prairie. And, seeing this landscape was very exciting and wide open with possibilities, it's not hard to put together in your imagination what could be here again. The idea—the mission—at American Prairie Reserve is to create the largest wildlife reserve so far ever created in the lower 48 states—open it up to the public, save it for future generations. We're creating the reserve out of existing intact, native prairie. There's very little of this left around the world where we could reassemble something like we're talking about here. The best thing about this is we don't have to buy it all. Much of the land that we are pulling together to be a part of this model already belongs to the public. What we're doing is gluing

these parcels together with pieces of private land, taking down the fences, bringing all the wildlife back, so we have intact prairie that looks like it has for thousands of years.[13]

The APR has already acquired thousands of acres of private land and leases to adjacent public grazing lands. However, the various patches of land that the APR now owns or controls are scattered throughout north-central Montana. The APR's land also borders land used for conservation and recreation, including the Charles M. Russell National Wildlife Refuge, Upper Missouri River Breaks National Monument, and the Fort Belknap Indian Reservation. APR's goal is to cooperate with federal, state, and tribal policymakers to ensure that the landscape is as seamless as possible, allowing wildlife to travel between these various land designations. If the APR succeeds in its goal, the entire complex of the APR's land and other conservation areas in the region will be a massive landscape. The APR prioritizes its private approach and has no intention to turn the land over to the federal or state government, but it wants to cooperate with federal and state land managers and wildlife managers to facilitate its overall goal.

In discussing the process of stitching together the full project, Fox has said, "It's a long game for a land acquisition. It's a probably even longer game for the restoration of habitats and species. This area was America's Serengeti—truly Americas Serengeti—with tens of thousands of bison, pronghorn, elk, deer, grizzly bears, wolves. [...] [The goal] is to have the ecosystem function fully as it once did."[14]

However, one of the major problems with restoring a landscape and allowing wildlife to flourish is that the wildlife can have negative externalities on the ranchers and farmers near the APR's land. For instance, ranchers who raise sheep or cattle see predators, such as wolves and bears, as a liability because each sheep or cow is part of the rancher's livelihood. Additionally, other herbivores like bison, elk, and deer can potentially harm the livelihood of nearby ranchers because wildlife can spread diseases, and when livestock die, that's a threat to a rancher's bottom line.

The APR's managers realized that their neighbors in Montana might not take so kindly to larger wildlife populations, so they have created incentive programs to mitigate the negative effects that neighbors might experience. To fund their incentive programs, APR started a subsidiary company called Wild Sky Beef, a for-profit company that sells grassfed beef to consumers throughout the United States. Profits from Wild Sky Beef are channeled in ways that can offset harm caused by the growing wildlife populations. By compensating neighbors who are harmed, wildlife shifts from being a detriment to a benefit.

If nearby ranchers choose to participate in the Wild Sky incentive program, they agree to certain conditions, such as not tilling their land, using nonlethal deterrents against predators, engaging in herbivore-friendly practices, installing wildlife-friendly fencing, removing fencing, or engaging in customized projects that promote biodiversity. As of 2024, each category of the Wild Sky program pays $0.75 per acre, per category, with total payments not exceeding $15,000. Additionally, the Cameras for Conservation program—a part of the Wild Sky program—pays nearby landowners to use trail cameras to capture photos or videos of wildlife on their land. The intention of Cameras for Conservation is to increase the tolerance for wildlife by providing a certain dollar amount for each species that is documented on private land. Currently, each property owner can earn up to $6,000 per year.[15]

Founder Sean Gerrity also said, "People are excited about the heritage of wildlife that they have read about, and they want to be a part of saving that as a part of American history. But, some people in the local area think that this may not be the best use of land out here—that perhaps this land should only be used for food production. So, we wanted to start something that benefits local people in the work that they're choosing to do."[16]

Daniel Kinka, a wildlife biologist, runs the Wild Sky program. He said, "Wild Sky ranchers are the best at what they do in terms of sustainable ranching."[17] For the camera program, Kinka said, "Get a coyote ... that's 25 bucks. A black bear comes behind it, that's 300 bucks. Four or five elk come behind it, that's 50 bucks. The biggest payouts are for wolves and grizzly bears—$500 per picture per camera per day."[18]

Dave Crasco described what his experience has been like being a rancher near APR and voluntarily participating in the Wild Sky incentive program:

Ranching has been in my family probably a hundred years. I'm a fifth-generation rancher here. You always hear the term that 'stuff gets in your DNA,' I guess, and that's probably what it is—it's just in my blood. My dad, he bought this place as a young man in his twenties. We've scattered his ashes here. I'll probably get scattered here, too, I suppose.

I think there's constantly threats facing ranchers. They're afraid of getting diseases from the buffalo. They're afraid of predators or land grab. It just seems like every time you turn around, the deck is stacked against you. So, when I heard about American Prairie Reserve, my first opinions were very negative. I was very much against them. It was like, "Go away. Leave us alone. That's not how we live here."

But, as I started educating myself, I got involved with Wild Sky Ranching, which is a branch of the APR. I saw it as an opportunity to make a few bucks—to utilize these conservation practices, whether it be wildlife friendly fencing or camera traps. Wild Sky sets up cameras, and then you get a picture of a mountain lion or bear, you get a couple hundred bucks. That adds up over the course of the year. That helps buy hay or part of a vet bill. It definitely helps my bottom line. And, so far, the camera traps have caught bears and mountain lions, and to my knowledge, I've not lost a single calf. So, I think we are living in harmony—living in balance—and wildlife is thriving and they're doing well.[19]

Lance Johnson, a fourth-generation rancher in Montana, also participates in the Wild Sky camera program. He said, "These ranches are so hard to make profitable that if you can figure out any way to supplement your income, then that's probably necessary right now."[20] Johnson also grazes cattle on a ranch owned by APR, so even though some other local ranchers may see APR as a threat, Johnson says that working with it helps ensure his family's future.[21]

Although APR's innovative approach has been relatively successful so far, it is still controversial. Connie French, a rancher who lives near APR, said that it feels to her—and many others like her—that for APR's vision to become a reality, ranchers and the ranching lifestyle must be displaced.[22] In APR's early days, APR leaders had used the phrasing "save the land" to refer to their goal, but that wording upset many local ranchers. APR's CEO Alison Fox has responded, "We are well aware that that word 'save' hit a nerve, and that was not at all our intention. Many of our ranching neighbors are committed to conservation. So, if I could pull back that word 'save,' I absolutely would."[23]

A few special interest groups in Montana have formed to oppose APR and its mission. One of these groups is the United Property Owners of Montana (UPOM), and its members see APR as a potential threat on several fronts, including the security of their property rights, the predictability of BLM leasing policies, and the longstanding culture of ranching/farming in Montana. Their slogan—found on signs throughout Montana—is "Save the Cowboy, Stop American Prairie Reserve."

One of the most effective ways that UPOM has found to oppose the APR is through BLM grazing permits. For the APR to maintain its public-land leases, it must graze an approved number of approved livestock, so the APR has been using cattle to maintain the leases. However, the APR's goal has been to restore bison populations, so it has been seeking permission to change from cattle to bison leases. In July 2022, the BLM issued a final decision that allows for bison on six of the APR's seven grazing allotments.[24] UPOM has taken the APR to court to stop the BLM from further allowing grazing permits, especially for bison, from going into effect.[25]

The APR is a fascinating case study because it shows how large-scale landscape conservation and restoration can happen in the private sphere, not just in the public sphere. Traditionally, environmentalists have thought that only governments were capable of such grand conservation projects, such as parks, wilderness areas, and wildlife refuges. But the APR shows how engaging in mutually beneficial exchanges and providing incentives to reticent neighbors can lead to positive environmental outcomes. In other words, private property and markets can lead to a cooperative, peaceful outcome for people who may have opposing ideas about landscape and wildlife conservation. The main caveat with this story is that politics is an arena of conflict, and that conflict is difficult to resolve because the nature of politics is zero-sum. The political fight over how BLM grazing leases are given out will continue to be a stumbling block for the APR's goals. Despite the bumps along the way, APR's approach to large-scale land conservation is an innovative and fascinating case study that may be replicable. A large-scale project like the APR may not be viable in all places, but it could be a useful and practical alternative in many cases.

14

Water Scarcity and Water Markets

With climate change, water is becoming scarcer in many places. The scarcity of water implies that one use must preclude another, meaning that conflicts will inevitably arise. Water can be allocated politically, through bureaucracies, regulations, and restrictions. Alternatively, water can be allocated using property rights, market prices, and voluntary contracts. Private property rights and entrepreneurship produce desirable results only when mutually beneficial exchange can occur. In deciding between these two approaches, Anderson and Leal argue that scholars, policymakers, and activists need to compare "whether politics or markets are more likely to produce efficient, amicable, and flexible solutions to the growing number of water conflicts."[1] When any resource, including water, becomes scarcer, some means needs to exist to incentivize consumers to use that resource more efficiently and to incentivize producers to find new ways of providing that resource.

Water markets can accomplish that. Water markets arise when people can hold private property rights to water and then trade those property rights willingly. In places across the globe—including Australia, the United States, Mozambique, Tanzania, Italy, Chile, and many more—policymakers have facilitated systems of tradable water rights, creating incentives for people to use water more efficiently by helping people conserve water when they otherwise wouldn't have.

Water markets are becoming more common around the world because markets tend to allocate water toward its highest economically valued uses. When people can trade the ownership of water, the intensity of how much somebody wants water can be considered. The people with weaker preferences will willingly trade their water rights to people with stronger preferences, and both parties are made better off through the trade. Like all prices, water prices are helpful because they signal the scarcity of water and its relative value compared to other competing uses. As such, prices lead people to trade their water from uses that are lower-valued to ones that are higher-valued. Market prices are also dynamic, which means they change with different conditions. During drought years when water is especially scarce, high prices will induce people to economize on their use of water, and during wet years, lower water prices will signal that people can use water more freely.[2] Since prices reflect the relative scarcity of goods, changes in water prices let farmers know when to improve water-use efficiency and invest in water-saving technologies. And, prices allow owners of water rights to see the opportunity costs of their choices. In other words, if set up correctly, water markets can resolve water shortages, encourage efficient use, and reduce government spending. Water markets also lower the transaction costs associated with allocating water. This combination of characteristics maximizes a society's total economic benefits.

When individuals own property rights to water and are alert to opportunities, they can trade those rights with one another so that water (literally and figuratively) flows to its most highly valued uses. By allowing the trading of water—even under conditions of extreme scarcity—people come to mutually beneficial arrangements peacefully. However, water markets will only facilitate peace if the individuals in the market perceive the property rights to be legitimate and justly acquired. If corruption is rampant, then a water market is less likely to fulfill its purpose and facilitate peace.

Like any other market, water markets have a demand side and a supply side, both of which can change. As climate change and population growth makes water scarcer in many places, markets can be used to deal with the risks of scarcity. Demand-side management can take many forms, such as educating people on how to decrease their water usage or how to make their water usage more efficient. Economic incentives could be used to lessen water demand, such as subsidizing the use of technologies that improve water efficiency. Market prices have the built-in advantage of discouraging overuse because people must bear the costs of the decision personally.[3]

On the supply side, more water can be provided through a number of means. More dams can be constructed, creating reservoirs that hold more water to be used later. However, dams have the tradeoff of disturbing river ecosystems, so this may not be a viable option in many cases. Additionally, desalination can extract freshwater from the ocean, providing usable water for domestic, agricultural, and industrial purposes. However, desalination requires a large amount of electricity, and landlocked areas are too far from the ocean to make desalination viable, so there are tradeoffs with this approach as well. For the purposes of water conservation, both demand and supply can be used to meet the competing claims on this precious resource.[4]

Water markets can take many different forms, but the core component is that water can be traded through voluntary buying and selling in some quantifiable form. Across the globe, three general types of water trading schemes have emerged. First, there are short-term or temporary transfers of water that is already allocated and available for immediate use. Second is a medium-term leasing of water allocations so that a user can plan ahead for a relatively long period of time. Third is a permanent transfer of water entitlements, which is a firm property right to water, either as a proportion of available water or a fixed quantity.[5]

Water markets take many different institutional forms. Across the world, people have traded water as informal arrangements between neighbors, which are governed by norms and social sanctions. Many countries have legally legitimized formal water markets, with governments overseeing and protecting the water rights of individuals.[6] One of the best examples of a highly functioning water market is in the Murray-Darling Basin (MDB) in Australia. The MDB water market has successfully allowed water users to trade water rights with one another. Economist and water policy expert Michael D. Young has argued, "Australia has defined its water entitlement and allocation arrangements in a manner that has made it possible to establish one of the world's most sophisticated water marketing systems."[7] Other scholars have called the MDB market "the most advanced water market in the world."[8] Australia's experience has demonstrated that water trading enables efficient and rapid adjustment to extreme water scarcity.[9]

The MDB water market was formally created in the 1980s and spans four Australian states and one territory, including Queensland, New South Wales, South Australia, Victoria, and the Australian Capital Territory. The MDB market is governed by both the Australian federal government and the relevant state governments under a joint agreement. At the federal level, an independent authority engages in basin-wide planning. The federal authority sets, monitors, and enforces water market rules, as well as monitoring, evaluating, and enforcing the Basin Plan. State agencies oversee how entitlements are issued and manage water use within agreed limits.[10]

The institutions that make the MDB water market work are a bit complicated, but they are an excellent case study to understand the strengths and weaknesses that water markets have for coping with the scarcity of water. In the MDB, two types of markets exist. One consists of permanent water rights, also called entitlements, which can be bought and sold. The other is a market of temporary water allocation.[11]

The water entitlement system in the MDB is based on a proportional system, meaning that each person is entitled to use a certain volume of water, depending on the availability of water that year. In contrast, the system in the

American West uses a prior appropriation system in which water rights are prioritized by seniority. If you are the least senior water-rights holder in a prior appropriation system, you are unlikely to receive your water during a drought year. The most senior water-rights holders are almost guaranteed to get their water, unless the rivers are completely dry. The Australian system, on the other hand, gives all entitlement holders the same status, with equal seniority. As such, trading entitlements in Australia have much lower transaction costs than the American West's system of prior appropriation because Australians don't need to check whether a trade would disadvantage other entitlement holders. Australia's system works well largely because there are relatively low costs for buying and selling entitlements, as well as temporary annual allocations.[12]

However, the Australian system didn't always work so well. Over time, Australia has improved its water market institutions when it saw that there were barriers in the way of trading water rights. Originally, after the MDB water market's creation in the 1980s, water entitlements were tied to agricultural land, meaning that only agricultural landholders could buy, sell, or own water entitlements. Since only agricultural irrigators could trade, the market was not very efficient because other people who valued water could not buy it legally.[13]

The original system of entitlements consisted of a "bundle" of sub-rights that specified how to use water, conditions about how it could be used, and so on. During this time, the sub-rights could not be unbundled, making it difficult to transfer water allocations from one location to another. Transaction costs were high, and entitlement holders reported that it took months to complete a trade. Under this system, the whole process of trading was convoluted. As economist and water policy expert Michael D. Young describes,

> The approach taken was to temporarily transfer the licence from one water user to another, then take the water off the licence and then, after the water had been taken from the licence, the licence was transferred back to the original owner. The process was slow and administratively complex. To

this day, this type of transfer is known as a temporary trade because the trade used to involve the temporary transfer of a licence from one person to another.[14]

However, political entrepreneurs saw the shortcomings of the old system, and they sought to facilitate a more efficient way of allocating water to its highest-valued uses. In the 1990s, policymakers first unbundled the land ownership from water ownership. Through a series of reforms, policymakers allowed people other than agricultural landholders to buy, sell, or own water entitlements, thus reducing the barriers to trade.[15] By unbundling water rights from land, landholders could obtain permission to irrigate an area of land without knowing where the water would come from, which increased investment security of formal water rights. Further reforms defined water licenses as shares and issued those shares in perpetuity. Even more reforms created separate bank-like water accounts, making it so that water could be allocated to each shareholder's account in proportion to the number of shares they held. Landholders who wanted to use water in an account would gain approval from a government official, who would deduct water from an account as it was used. While not directly related to the reforms, policymakers implemented other parallel reforms to system-wide planning processes that gave irrigators more confidence, predictability, and stability.[16]

Even though these reforms are complex and perhaps confusing to people encountering them for the first time, they created prime conditions for a well-functioning water market to rapidly develop, with much lower transaction costs than before. A highly efficient system of water trading emerged.[17] By unbundling water entitlements, Australian policymakers increased the speed of trading of water allocations, and in recent years, most trades take less than two days, as opposed to months.[18] A variety of investors who did not own land began participating in the market, and the volume of water trading in the MDB increased enormously. Like other markets, entrepreneurs saw opportunities to supplement the MDB water market with derivative products, including forward contracts, entitlement leasing, and carryover capacity leasing.[19] Each of these derivative products made it easier to trade water entitlements at lower costs and with a wider variety of services to meet the diverse needs of water users in the basin.

Australia's experience with water markets is living proof of economic logic. When people were free to trade water with minimal barriers, market prices for water emerged. The prices signaled to irrigators the relative scarcity of water and the opportunity costs of using water in a particular way. In an empirical study, scholars Zhao, Ancev, and Vervoort found that an increase in rainfall increases irrigation water supply and curtails irrigation demand, leading to a decrease in prices. They conclude that the "price mechanism in the MDB water market functions well in terms of signaling the level of water scarcity/supply in the region across multiple important trading zones."[20]

For the most part, the MDB water market appears to have generally been beneficial for nearly everyone who participates in it. Academic research indicates that, in the water market, the vast majority of participants were made better off because the people who chose to sell their water entitlements could make more money from selling water than by using it. Evidence also suggests that entitlement holders chose to sell water to finance investment in more efficient irrigation technology. However, in recent years, the MDB has suffered with over-allocation problems, so the Australian government has been buying water entitlements for the environment from irrigators willing to sell some or all of their water to them. As water leaves a district for environmental or irrigation purposes, the unit costs of supplying water to remaining irrigators can increase, which may make it hard for less wealthy irrigators to continue.[21]

Despite the MDB water market's general successes, including the more efficient and effective approaches to reallocate water, critics have voiced concerns. They argue that markets commodify water to benefit the wealthy and powerful at the expense of the most vulnerable. When large income inequalities exist, people may develop negative perceptions of markets as an acceptable means of reallocating water.[22] Additionally, some scholars have acknowledged that the Australian system has a number of shortcomings, such as limited accuracy in water accounting and a lack of accounting for wider social impacts.[23]

Many more examples of water markets exist across the world, but some of them have not been as successful as Australia's. For example, Chile has had some form of water markets since the early 1980s, but many conflicts—sometimes violent ones—have erupted over how water is allocated and how water rights are managed.[24] On the other hand, in India and Pakistan, water markets have "stopped fighting among farmers and engendered cooperation. Markets replaced conflict," as Shah and Maitra have argued.[25] Water markets may not perfectly allocate water or forestall violence, but the question is whether water markets function better on those margins than relevant alternatives. Evidence suggests that the political allocation of water—the opposite of a water market—has been directly or indirectly responsible for armed conflicts in Bangladesh, Tajikistan, Malaysia, Yugoslavia, Angola, East Timor, Namibia, Botswana, Zambia, Ecuador, and Peru in recent decades.[26]

Although the MDB example shows how water markets can function relatively well, there are other types of institutions that can be used to address water scarcity issues, such as block pricing and water trusts. Block pricing is a way to charge households and businesses for water in different increments so that people who consume more water are charged higher prices. In other words, water becomes more expensive as more water is consumed. In the United States, some of the largest and fastest growing cities are in California and the Southwest—an area that has been suffering with a decades-long drought—but these areas often have the cheapest municipal water prices. To improve municipal water use, block pricing incentivizes people to decrease their use of water willingly. For example, the first thousand gallons of water is the cheapest and will cover basic necessities. The next thousand gallons will have a higher price, and the next thousand gallons will have an even higher price. This pricing structure gives people an incentive to use water more efficiently than they otherwise would have, but it also doesn't punish people for using water for necessities, such as drinking and bathing.[27]

For example, Los Angeles—a notoriously dry city—uses a block pricing system to incentivize prudent water usage. The system has four blocks: tier 1 for basic indoor use, tier 2 for efficient drought resistant outdoors water use, tier 3 for above average outdoor use, and tier 4 for excessive use. The blocks also depend on a property's lot size, season, and temperature zone. There are five categories for lot sizes (based on square footage), two categories for seasons (summer and winter), and three temperature zones (low, medium, and high).[28] While this system may seem overwhelmingly complex at first, the purpose is to ensure that people economize on their water usage when it is most prudent to do so. For example, someone who excessively uses water on the largest lot during summer in a high temperature zone will pay the highest prices for water.

Another institution to conserve water is a water trust—a system in which a person or organization can acquire water rights and then dedicate that water to environmental conservation purposes. These purposes might include leaving the water in a stream or river for ecosystem health, or it can be used to preserve wetlands. One important example of a water trust is related to Utah's Great Salt Lake, which has been slowly shrinking since it reached its largest size in the 1980s. Robert Gillies, Utah's state climatologist, has explained,

> What has happened since the pioneers first came here is they started diverting water from the rivers that fed the Great Salt Lake, so the amount of water that is being diverted from these rivers has increased exponentially and is going to do so into the future because our population is booming.

But then there's a secondary factor, and that factor is climate change—because climate change here in northern Utah is changing the whole hydro climate of the region.[29]

While the Great Salt Lake has always fluctuated in size, Utah's fast-growing population and climate change caused the Great Salt Lake to reach its lowest level in 2022. This is especially important because the lakebed contains many hazardous materials, including toxins like mercury, arsenic, and selenium, which can blow directly into Salt Lake City and its suburbs. In response to the record-breaking low levels of the lake, the Utah State Legislature unanimously authorized $40 million for the Great Salt Lake Watershed Enhancement Program, which set up a water trust to enhance water quantity and water quality for the Great Salt Lake and its wetlands.[30] In 2023, the Great Salt Lake Watershed Enhancement Trust awarded a total of $8,525,343 to eight projects from local, state, federal, and nongovernmental entities to protect and restore the wetlands and hydrology of the Great Salt Lake.[31]

The Great Salt Lake is a helpful example for studying water issues more broadly. As environmental scholar Katie Wright said, "At the end of the day, without exception, it's all about incentives. In this case of the Great Salt Lake, it's about answering the question: How can we devise policies that recognize individual motivations and yet reward everyone coming together to achieve a common goal? Nobody wants the lake to go away."[32] In addition to keeping water in the lake to avoid the toxic materials in the lakebed, the Great Salt Lake is home to one of the largest wetland complexes in the western United States, providing a critical stopover and breeding ground for more than ten million birds annually.

In the western United States, including Utah, the prior appropriation system of water rights is governed by laws that specify how water rights can be used. Utah, like several other western states, specifies that water must be put to a "beneficial use," which has been legally defined as consumptive uses in farms, homes, and industry, but not habitat for fish and wildlife. In other words, water that has been left instream for environmental conservation purposes is not being put to a beneficial use, meaning that the water has been "wasted," so the water is made available for a more junior water-rights holder who will use it for a beneficial use.

Tim Hawkes, who spent more than a decade as Utah state director for the environmental nonprofit Trout Unlimited, wanted to reform Utah's beneficial-use laws. Beginning in the early 2000s, his goal was to reform Utah water laws to make it easier for conservationists, including nonprofit groups like Trout Unlimited, to lease water from farmers. By leasing the water, they could leave it instream so that aquatic ecosystems were healthier, while also making fishing more enjoyable for anglers. Utah had already slightly modified its water laws in the 1980s, but it only allowed Utah's Division of Wildlife Resources and Division of State Parks to acquire and hold instream rights. These agencies, however, rarely used their authority to do so. Hawkes wanted to expand these reforms so that private individuals and other groups could lease or buy water rights for conservation purposes.

Hawkes had a difficult time convincing both state lawmakers and farmers to consider a change to Utah's water laws. Members of the Utah Farm Bureau—many of whom owned senior water rights—were skeptical of Hawkes's claim that changing Utah's water laws could potentially benefit them. However, Montana had changed its water laws a few years earlier so that people could lease water rights to leave water instream. Hawkes invited members of the Utah Farm Bureau on a trip to Montana to meet with farmers and ranchers who had already experienced a system in which people could have private leases for instream flows. Many members of the Utah Farm Bureau were persuaded by the Montana ranchers that a change to Utah's water laws could be beneficial.

Hawkes drafted a bill to present to the Utah legislature, with many restrictions that would assuage much of the remaining skepticism of farmers and ranchers in Utah. Some of the restrictions in this early draft included a ten-year "sunset" when the law would expire, a ten-year time limit on leases, restrictions on how far downstream protected flows could go, loss of seniority for leased instream rights, and restrictions on who could hold the rights, among others. Over three years, the bill was amended, growing from a few sentences to over four hundred lines. The Utah Farm Bureau gave its official support, and in 2007, a Senate committee unanimously approved the bill. In the House, however, the bill barely passed in a committee vote. One of the representatives who opposed the bill was afraid that voluntary leases could be turned into involuntary ones through the application of federal law. In other words, he feared that a policy experiment might accidentally become permanent due to provisions in the Endangered Species Act. When the bill went to the full House for a vote, it failed by only two votes.

The battle to change Utah's water-rights institutions went on for years, with some successes. In 2008, a bill passed the Utah Legislature that allowed water leasing, as long as a lease was accompanied by a Candidate Conservation Agreement with Assurances (CCAA) with the US Fish and Wildlife Service. Under the Endangered Species Act, a CCAA shields participants in the leasing program from punitive measures if the Fish and Wildlife Service later decided to list the trout species that the lease protected. However, bureaucratic hurdles meant that very few CCAAs were granted, so very few leases were processed. In 2013, Hawkes helped the legislature create a second path to water leasing that didn't require a CCAA. A lease could proceed, as long as the lessee agreed in the contract to assume any liability under the Endangered Species Act and protect the owner of the water right against that risk. Between 2013 and 2018, Trout Unlimited secured several modest leases around Utah.

In 2015, Hawkes was elected to the Utah House of Representatives, and in 2019, he introduced a bill that made the private leasing program permanent by removing the ten-year sunset provision. This bill was supported by the Utah Farm Bureau, and the legislature passed the bill unanimously. In 2022, the Utah Legislature passed a new bill that made the instream flow program easier to use and administer by allowing any private person or entity to lease water for instream flows. This bill also allowed a water-right holder to convert an existing right to an instream flow on a temporary basis without involving any other party. The Division of Forestry, Fire, and State Lands could also acquire and hold instream flow rights to benefit state sovereign lands, including the Great Salt Lake. Lastly, the new bill removed the CCAA requirement and many other restrictions. The 2022 reforms to the water leasing program were important because they made instream flows more like traditional, consumptive rights, meaning that it would be even easier for private individuals, nonprofits, and state agencies to lease water rights for environmental purposes.[33]

This episode in Utah is instructive for several reasons. It shows how environmental problems are, at their core, caused by property rights and the institutions that govern property rights. Sometimes, legal institutions make it difficult for people to engage in trades that might benefit environmental outcomes, such as leasing water to leave instream for fish populations. Political entrepreneurs, like Tim Hawkes, can use their expertise to find alternative institutions that allow people to discover ways of improving the environment, but implementing institutional reforms often requires time and patience. When institutions are reformed in ways that remove barriers to trade, there can be win-win scenarios. Environmental groups won because they were allowed to lease water rights that improve habitat for fish. Owners of water rights, like farmers and ranchers, made money by leasing water rights to the environmental groups who were willing to pay. Additionally, a polycentric approach to the institutions that govern water rights was important for experimentation and learning. People in Utah were able to observe what Montana had done with their water leasing program and saw that it was a workable system. Utah policymakers then tailored their water leasing program to their unique circumstances.

In summary, many different institutions exist that can help with the allocation and conservation of water. One way is through government bureaucracies, and another way is by creating property rights and allowing people to trade in markets. Either approach—bureaucratic or market—can allocate or conserve water, but policymakers and citizens need to evaluate the different knowledge- and incentive-related characteristics of both approaches. Although the mechanisms of the market work in an idealized theory, those mechanisms do not always play out as effectively as assumed. Similarly, we can imagine how an ideal form of bureaucratic governance could allocate and conserve water, but the real-world practice of politics and public administration can often lead to ineffective and inefficient outcomes. A relatively well-functioning water market may be the correct choice for many scenarios because it allows for positive-sum, mutually beneficial exchange, while also allowing water to flow to its most highly valued uses, both literally and figuratively. Water markets have prices, and like all prices, buyers and sellers can use the knowledge contained in those prices to realize the scarcity of water at any time and also assess the opportunity costs of their choices. Political allocations, on the other hand, are zero sum, and may often become negative sum through the rent-seeking process. When water is especially scarce, negative-sum games have higher stakes, which spark violent conflict as individuals fight for access to water.

A water market approach does not necessarily remove any need for public policy or government oversight. There are many potential ways to institute a water market that provide the economic and efficiency-related benefits of markets, while also allowing for public policies to ensure access to water based on other normative concerns, such as the unequal access or distribution of resources. However, government overseers of water markets can sometimes unintentionally impose barriers or cause distortions to water markets, or they may engage in corruption or cronyism that undermines the benefits of a market approach. Currently, many public policies stand in the way of finding creative ways to solve water-scarcity issues. By removing barriers to trade, people can act more entrepreneurially in the way they use or conserve water.

15

Climate Change

Human-caused climate change is one of the most pressing social and environmental issues of modernity. People across the globe are beginning to feel the difficulties associated with climate change. In a 2022 survey of German residents, 85 percent of respondents thought that human-induced climate change exists, and 83 percent thought that it has an impact on human health.[1] In a 2023 survey of Americans, a majority said that they see climate change as a major threat, with 37 percent saying that addressing climate change should be a top priority for the President and Congress, and another 34 percent say it's an important but lower priority.[2] In a 2020 survey, 63 percent of respondents in Japan said that climate change is affecting where they live, either a great deal or somewhat.[3]

Even with growing anxiety over climate change, it isn't just one problem. It's a web of interconnected problems at various scales and scopes. Both the causes and consequences are highly complex phenomena. The causes are largely due to greenhouse gas emissions from transportation, agriculture, and industry. Since the beginning of the Industrial Revolution, humans have been emitting large amounts of greenhouse gases into the atmosphere by burning fossil fuels. Of course, fossil fuels are beneficial because they are very energy-rich, providing large amounts of energy reliability. However, the tradeoff with using such a handy energy source is that carbon dioxide is a waste product, and as carbon dioxide levels increase in the atmosphere, it acts as a blanket, trapping in heat. Additionally, our growing wealth has made our population increase, but with more people means more mouths to feed. Agriculture, both the growing of fertilizer-intensive crops and the raising of livestock, has also contributed to the emissions of greenhouse gases, especially methane.

The consequences include many problems arising simultaneously and to different degrees in different places. Many places are becoming hotter. Some places are becoming drier, while others are becoming wetter. Oceans are acidifying, and sea levels are rising. Storms are becoming more intense, and many glaciers are retreating.[4]

Any reasonable person might ask, "What are we to do in the midst of all this complexity?" There is no simple or straightforward answer. Some prominent scholars and activists have called for top-down, centralized approaches, similar to the war-time planning that was used in World War II or communist economies.[5] Other scholars and activists disagree with the top-down, centralized approach, and those working in the tradition of Elinor Ostrom see potential solutions in a more decentralized, polycentric approach. Ostrom argued that top-down, once-size-fits-all policies are unlikely to solve the climate change problem, and in some ways, they may make it worse.

Although many policymakers are experts, they are not omniscient. The processes of effective mitigation and adaptation are countless and complex, leading to knowledge problems on many fronts. It's not always clear what kind of regulations are best, or what the best scale or scope of regulation should be. It's hard to know the opportunity costs of policy decisions. Policymakers can't know exactly what kinds of negative unintended consequences or regressive effects will emerge from well-intended policies.

Besides innumerable knowledge problems, centralized policymaking also leads to many incentive problems, as described in public choice theory (see Chapter 9). When political power is concentrated, policymakers have a strong incentive to wield that power in ways that benefit them and their supporters at the expense of others. The processes of rent-seeking lead to government-granted privileges, creating a system of concentrated benefits and dispersed cost that are wasteful from a social perspective.[6]

To avoid the worst of the knowledge problems and incentive problems that accompany centralized power, Ostrom argued for a polycentric approach for coping with climate change, which she saw as the most viable way forward.[7] As discussed in Chapter 10, when decision-making power is arranged in a polycentric way—meaning it is spread out rather than centralized—solutions to complex problems are more likely to emerge. The daunting and dire problems associated with climate change lead many intelligent people to assume that national governments or the United Nations must necessarily be at the core of solving a global-sized problem.

However, the logic of polycentric governance pushes back against that assumption. As Ostrom argued, global problems like climate change do not have to rely solely on global-level actions: "[G]iven the slowness and conflict involved in achieving a global solution, recognizing the potential of building even more effective ways of reducing energy use at multiple levels is an important step forward."[8] Ostrom saw that polycentric approaches to climate change need to be expanded and bolstered, and decision-makers should resist the temptation to centralize any mitigation and adaptation activities.

A polycentric approach to climate change has several advantages. Competition among different entities gives them a stronger incentive to work for society's benefit. Cooperation among those same entities allows them to work together and combine knowledge at different scales and scopes, creating opportunities for synergy. When decisions are made at lower levels, people are more likely to see them as legitimate, which in turn makes coproduction more likely. Heterogeneity becomes an asset because different preferences and worldviews lead to different approaches, thus leading to more opportunities for experimentation and learning. When some of these approaches fail to reach their goal, the whole system is more resilient than if a single, centralized goal were to fail. The combination of many different approaches can yield desired outcomes without requiring central planning.[9]

At the biggest scale, a polycentric approach engaging in effective mitigation of climate change, as well as adapting to it, will require actions from various governments, market firms, civil society groups, and academia. The combination of these different spheres of human action makes up a "meta-polycentric order" of smaller, overlapping polycentric systems. For instance, most governments are internally polycentric because power is split between different levels and different branches. Markets are polycentric because many firms compete against one another in some cases, but they also cooperate with one another in other cases. Civil society has many different clubs, associations, organizations, and groups that each have different goals and methods for achieving those goals. Academia engages in scientific contestation to discover and test new truths.

Markets—as a clear example of a polycentric system within a broader meta-polycentric order—are one of the most important sources of mitigating and adapting to climate change. A wide variety of innovations have arisen from the competitive pressures of the market, sometimes with pressures from public policies. Here are just a few examples of market-driven improvement:

- Internal combustion engines have continued to become more fuel efficient and emit fewer pollutants.[10]
- Electric vehicles are quickly improving, with longer battery life and higher efficiency.[11]
- A combination of government subsidies and market competition has made solar panels more affordable, leading to broader adoption in both residential and commercial settings, although some lower-income communities are lagging behind.[12]

- New innovations in wind turbine technology have become more efficient at producing electricity, making wind energy an increasingly viable energy source.[13]

- Improvements in lithium-ion batteries have enabled better energy storage, which is especially important for electric vehicles and renewable sources.[14]

- Other forms of energy storage technologies, including pumped hydro storage, compressed air energy storage, and advanced grid-scale batteries, have increasingly helped to improve electrical grid reliability.[15]

- New technologies have led to the widespread adoption of smart thermostats and HVAC systems, which optimize energy use in domestic and commercial spaces, thus reducing energy consumption and emissions.[16]

- Innovations in various kinds of building materials have caused construction to become "greener," reducing carbon footprints across the globe.[17]

- New technologies and techniques in farming, known as precision agriculture, have allowed food production to use less water, require fewer fertilizers/pesticides, and produce larger yields, all with fewer environmental impacts.[18]

- Carbon-capture-and-storage technologies, which have been driven largely by market advancements, are becoming increasingly viable in an economic sense. It is possible that these storage technologies can reduce the amount of carbon dioxide that has already been emitted into the atmosphere, addressing a key cause of climate change.[19]

One fascinating technology emerging from markets is the Climeworks' Orca carbon-capture facility in Hellisheiði, Iceland. Orca uses giant collectors to filter carbon dioxide from the atmosphere. As Bryndís Nielsen, a Climeworks spokesperson, explained: "The filter captures the CO_2 molecules while releasing the cleaner air back into the atmosphere. And, once the filters are full, you close off the compartment that the filter is in, and you apply heat to around 100°C, which releases the CO_2 that we can then capture and collect." The heat is produced renewably from a geothermal power plant, and it currently collects about 4,000 tons of CO_2 per year, so far. As Neilsen explained, "[Four thousand tons per year] is not a lot. I mean the globe emits around 36 billion tonnes of CO_2 per year, but this is a new technology. We went from capturing milligrams to kilograms to tons to [4,000] tons, so we are on our way to a huge and massive scale-up."[20]

Another company called Carbfix is an Icelandic company working on large-scale carbon capture. Ólafur Teitur Guðnason, the Head of Communications and External Relations at Carbfix, described how it works to capture carbon. They inject CO_2-charged water "down into the earth where the CO_2 from the water is mineralized in less than two years." Since Iceland is largely made of basalt, it can inject the water into the porous rock. The water then "goes through these holes, and the CO_2 in the water dissolves metals in the rock and forms a mineral called calcite." Eventually, the CO_2 is "turned into stone, filling up the holes in rock."[21] Through this process, the CO_2 can stay in the rock essentially forever. However, Guðnason says we can't rely completely on carbon capture: "You cannot use this technology as an excuse for 'business as usual.' It can only be an addition to all of our other efforts in reducing emissions, reducing the dependency on fossil fuels, changing over to greener sources of energy. We need to do all of those things."[22]

Collaborations between academia and market firms have also produced promising discoveries. Ermias Kebreab is the associate dean for global engagement in the College of Agricultural and Environmental Sciences at the University of California, Davis, as well as director of the World Food Center. His research has made new discoveries about reducing methane—a far more potent greenhouse gas than carbon dioxide—in beef and dairy production. A few researchers have discovered that seaweed contains chemicals that directly slow or stop microbes in a cow's gut from forming methane, and it does this without interfering with food digestion. Kebreab had read an academic article on how adding seaweed to chopped grass drastically limited methane emissions from cows, but this was in a lab setting. Kebreab didn't doubt the lab results necessarily, but he was a bit skeptical because many lab results don't always translate to real-world settings. He and his team chose to test how adding seaweed to the diet of live animals would affect methane emissions. Kebreab collaborated with many people—including entrepreneur Joan Salwen, scholars from James Cook University, and researchers from Australia's Commonwealth Scientific and Industrial Research Organization—to conduct an experiment.

Their experiment was the first ever in dairy cattle, which meant they were starting from scratch. The research team didn't know how much seaweed to feed the cattle, so they started with about 60 grams of seaweed per 25 kilograms of feed per day. They also increased the amount of seaweed up to 250 grams per 25 kilograms of feed to see how emissions changed with the ratio of seaweed. One of the researchers oversaw "trapping" the cows' burps to measure the volume of their methane emissions. In just the first experiment, the researchers found that adding seaweed reduced methane emissions by up to 67 percent. The research team was so astounded by the results that they thought that the equipment had malfunctioned, but after checking everything, the results were legitimate.

Even with the initial success, more questions immediately arose. They wondered whether the microbes in the cows' gut would become used to the seaweed and start producing methane over time. They also wondered whether seaweed that was stored for a long period of time would be as effective as freshly harvested seaweed. Maybe cows would not like the taste of seaweed in their feed, limiting the efficacy of adding it. Or, maybe seaweed would have other unexpected effects, like harming the cows' health or reducing the milk and meat production.

The research team started another trial over a five-month period, and their worries were largely unfounded. Seaweed that was harvested three years prior still reduced emissions by over 80 percent. Other colleagues in Australia reported up to a 98 percent reduction in methane in a similar trial. More trials found that adding seaweed did not affect the health of the cattle, and it did not affect the quality of the meat. They put together a panel of 112 people, some of whom tasted steak made from the steers offered seaweed, and some of whom tasted steak from a control group. The panel did not taste any difference in the steaks, and another analysis showed that the nutritional quality of both kinds of meat was the same. All in all, Kebreab's research shows how new discoveries and innovations can lead to a triple-win scenario: a win for the environment, the farmers, and the consumers.[23]

More research is being done to figure out whether supplementing cattle feed with seaweed is a viable option. Of course, growing, harvesting, processing, storing, and transporting seaweed will produce greenhouse gas emissions, so it is important to find out whether the reductions in methane from feeding cattle seaweed will be offset by any emissions from producing the seaweed itself.[24] Other factors, like the seaweed species that is harvested, the geographical location of cattle farms, and the rates at which seaweed is incorporated into cattle feed will affect methane suppression.[25]

Perhaps the biggest issue is the cost to farmers. In a 2023 study of organic dairy farmers in Maine, many of them had already heard about the climate-related benefits of incorporating seaweed into their cows' diet. However, their receptiveness to implement seaweed in dairy diets depends largely on the costs of obtaining the seaweed. The farmers expressed other concerns as well, including co-benefits in terms of cattle health and performance, government policies and subsidies, and the ease of integration into existing feeding management practices.[26] In short, seaweed is no panacea, but it may be one element in a broader polycentric arrangement of approaches. As the market for seaweed develops, it seems likely that quantity will go up, and the price will go down, as with many other goods and services.

Other academics are studying and debating new means of mitigation. Ulrike Niemeier, of the Max Planck Institute for Meteorology, has been researching other solutions to climate change, such as forms of atmospheric engineering. This might include using aircraft to introduce sulfur into the stratosphere, which would help reflect sunlight, leading to cooler temperatures. However, as Niemeier warns,

> If we brought sulfur into the atmosphere, it would be globally distributed in all directions by air currents. It would affect everybody. We have to see what the impact of a sulfur emission up there would be. What kind of climate effect would it have? Maybe there would be side effects we can't easily foresee. It could change rainy and dry seasons regionally, with consequences for farmers. It's also not clear how much sulfur would be needed. Sulfur has a lifetime of about one year, so that means you would have to bring up large amounts continuously, or at least once, twice, three times a year to get a fairly constant layer. If you stop the whole thing, the sulfur would be gone again in two years at most, and the Earth's temperatures would rise rapidly again.
>
> It should really only be used as a last resort. But if we don't reduce emissions, I think life on Earth will become very difficult in some regions, which could very well tip the public mood in favor of such methods. Of course, this would mean creating another artificial solution on top of artificial climate change, which is not such a nice idea either.[27]

A polycentric approach to climate change involves government action at different scales and scopes. One such policy is California's Cap-and-Trade Program for greenhouse gas emissions. The program took effect in 2013, and each year since, the California Air Resources Board has issued fewer allowances, leading to declines in emissions over time. The program supplements existing rules and regulations for air quality and toxic air pollutants.[28] California's Cap-and-Trade Program is mandatory for facilities that emit more than 25,000 metric tons of carbon dioxide equivalent per year, which includes electricity generators, as well as refineries, cement production facilities, oil and gas production facilities, glass manufacturing facilities, and food processing plants, among many others.[29]

While California's system has not been flawless, it has generally worked well to reduce greenhouse gas emissions since it was put in place. Economists Christian Lessmann and Niklas Kramer found in 2024 that California's program has decreased the emissions from the state's power sector by 6 percent annually, compared to a "synthetic control" they created in statistical modeling. Their research found, however, that the cap-and-trade system hadn't done much to reduce emissions in the industrial sector, increasing on average by less than 1 percent annually, relative to their statistically created synthetic control. They conclude that their findings, among evidence from other researchers that they reviewed, suggested that California's carbon trading system has had limited success in reducing emissions, particularly in the power sector, but it has been less effective in other sectors.[30]

Economists Danae Hernandez-Cortes and Kyle Meng found that California's Cap-and-Trade Program was successful at reducing a number of air pollutants between 2012 and 2017. On average, the annual decrease in greenhouse gas emissions was 9 percent. Fine particulate matter less than 2.5 microns in diameter ($PM_{2.5}$) decreased by 5 percent, and fine particulate matter less than 10 microns in diameter (PM_{10}) decreased by 4 percent. Various forms of nitrogen oxides decreased by 3 percent on average per year.[31]

However, cap-and-trade systems, like California's, may be relatively successful at reducing pollution, but they also have tradeoffs. A market-based environmental policy like a cap-and-trade is designed specifically for "allocative efficiency," meaning that resources flow to people in a way that maximizes overall satisfaction and welfare in a society (see Chapter 11 for more detail). But there are concerns about "environmental justice," meaning that communities with lower income who are otherwise marginalized will systematically experience higher pollution concentrations than other communities. Since cap-and-trade systems allow high-cost polluters to continue to pollute, as long as they buy permits from lower-cost polluters, pollution can become geographically concentrated. There are concerns with normative considerations, like justice, if that concentration affects lower-income groups or marginalized communities at significantly higher rates. Therefore, many economists, philosophers, and policymakers have wrestled with the apparent (or assumed) equity-efficiency tradeoff of cap-and-trade systems, as well as other market-like public policies.[32]

Hernandez-Cortes and Meng, however, find that California's Cap-and-Trade Program may have actually helped, or at least not exacerbated, environmental justice concerns, while also successfully reducing emissions overall. Before the program was put in place, they found that the "environmental-justice gap" was growing in California between 2008 and 2012. Since the program started, the previously widening environmental-justice gaps have been narrowing. Based on their calculations, Hernandez-Cortes and Meng find that between 2012 and 2017, California's Cap-and-Trade Program has reduced a number of environmental-justice gaps: $PM_{2.5}$ has decreased the gap by 7 percent, PM_{10} has decreased it by 6 percent, and nitrogen oxides have decreased it by 10 percent annually. This isn't to say that there are no environmental-justice gaps left, however. While it appears that the program has narrowed the gaps, they still remain. By 2017, gaps returned to roughly the 2008 levels.[33] Of course, concerns about environmental justice may come up with other cap-and-trade systems, so scholars and policymakers should evaluate this on a case-by-case basis.

While California's cap-and-trade system is completely within one state, a multi-state cap-and-trade system exists in the northeastern United States called the Regional Greenhouse Gas Initiative (RGGI). Starting in 2009, the RGGI was the first mandatory market-based cap-and-trade program in the United States. As of 2023, it consisted of Connecticut, Delaware, Maine, Maryland, Massachusetts, New Hampshire, New Jersey, New York, Rhode Island, and Vermont. The state governments in the region established a cap on carbon dioxide emissions from power plants and issued a limited number of tradeable allowances through quarterly auctions. State governments invest the proceeds from the auctions in energy efficiency, renewable energy, and other consumer benefit programs.[34]

Overall, the RGGI's approach has been successful, especially in its early years. Between 2009 and 2014, carbon emissions in the states participating in the RGGI dropped 35 percent, which is significantly higher than the 12 percent reductions in non-RGGI states. Additionally, participation in the RGGI didn't seem to harm the state economies. Participating states in the RGGI saw their economies grow 21.2 percent, compared to 18.2 percent in non-RGGI states.[35]

Despite the successes, the RGGI has tradeoffs, with one of the main ones being the displacement—or "leakage"—of greenhouse gas emissions to nearby states that don't participate. Economist Jingchi Yan finds that the RGGI has led to the phase-out of coal and natural gas power plants in participating states. In fact, since the program began, electricity generation from coal has decreased 73 percent, and electricity generation from natural gas has decreased 30 percent within regulated states. However, in nearby states that don't participate in the RGGI, natural-gas consumption has increased 237 percent, but coal consumption has decreased 7 percent. As a result, reductions in participating states have been partially offset by increases in nearby states. Overall, participating states reduced their carbon dioxide emissions by 4.8 million tons annually, but non-participating states increased carbon dioxide emissions by 3.5 million tons, yielding a net reduction of carbon dioxide emissions of 1.3 million tons per year.[36]

Economists Harrison Fell and Peter Maniloff also found that emissions have been somewhat displaced from RGGI states to nearby Pennsylvania and Ohio that do not participate in the RGGI but can transmit electricity to RGGI states. In their calculations, states in the RGGI have decreased CO_2 emissions at about 8.8 million tons annually, but surrounding states have increased theirs by about 4.5 million tons annually, resulting in an aggregate decrease of 4.3 million tons per year. Although the RGGI still yields a net reduction, the leakage of emissions to other places is an undeniable tradeoff. In other cases, such leakage might be worse, and it may actually lead to higher emissions overall if energy production moves to more emissions-intensive areas.[37] Again, scholars and policymakers need to evaluate this potential drawback on a case-by-case basis.

In recent years, policymakers have realized the need to include many more types of stakeholders in the design of environmental policies, particularly cap-and-trade systems. In 2016, the RGGI held an official program review to update the governance structure. Between November 2015 to September 2017, the RGGI held a series of nine meetings and webinars, each followed by written feedback periods. Interested stakeholders—energy companies, private citizens, environmental groups, clean energy advocates, governmental entities, other business interests, academic scholars, and carbon market specialists—attended these sessions and provided input. These stakeholders provided suggestions but were not true decision-makers in the process.

As expected, different combinations of stakeholders grouped together based on their preferences, forming coalitions over the way the reforms should be designed. After all the feedback from the various groups and coalitions was gathered, the RGGI's Board used that feedback to create the "2017 Model Rule," which contained the new standards and rules agreed to by each member state. However, the Board did not provide any formal response to stakeholder input or an explanation of the updates to the policy design, making it difficult to know how stakeholder feedback was specifically incorporated. As environmental policy scholars Maxwell Dorman, Aaron Strong, and Nicola Ulibarri argue, "The black boxing of this mitigation policy redesign process leaves it unclear whether stakeholder involvement in this process was truly formative or merely tokenistic. The updates made in the RGGI policy redesign process resulted in a climate mitigation policy with 'increased regulation intensity.' Although this aligns with much of the input from stakeholders supporting 'increased regulation,' and the degree to which it does not accord with input from 'decreased regulation/status quo' stakeholders, the black boxed decision-making process means we cannot be sure whether the Board would have made these same changes without stakeholder input."[38]

There is not one correct way to gather stakeholder input for policy design or redesign, but due to the dispersed nature of knowledge, policymakers should consider different ways of gathering knowledge that they may not have access to. The RGGI's 2016 Program Review can serve as a learning opportunity for ways in which to gather knowledge from a variety of sources, sort through that knowledge, evaluate the preferences and incentives of each contributor, and evaluate the tradeoffs of each potential decision. Insights from public choice economics are also helpful because of the incentive for special interest groups to attempt to influence decisions in ways that benefit themselves at the expense of others. In short, a polycentric approach to policymaking for climate change must involve a wide range of viewpoints, preferences, and knowledge sources, otherwise policymakers risk creating ineffective or inefficient policies, or accidentally sparking negative unintended consequences.

In a meta-polycentric order, nonprofit and advocacy groups play an important role in ensuring that markets and governments have checks and balances on their actions. One recent example of this is with Pennsylvania and its (potential) participation in the RGGI. In 2019, then-Governor Tom Wolf announced Pennsylvania's intent to join the RGGI. In July 2022, Pennsylvania began participating in the RGGI, but soon after, the Commonwealth Court of Pennsylvania issued an injunction that stopped the state's participation.

On November 1, 2023, the Commonwealth Court of Pennsylvania ruled that RGGI participation violates the state constitution. Later that month, the Governor's office appealed this decision to the state Supreme Court.

Since then, a number of nonprofit organizations and advocacy groups—including PennFuture, the Clean Air Council, the Sierra Club, and the Environmental Defense Fund—have filed briefs with the Pennsylvania Supreme Court to reverse the lower court's decisions, allowing the state to rejoin the RGGI.[39] The Executive Director of Environmental Defense Fund Amanda Leland wrote, "We agree with the Governor [Shapiro] that the state's Environmental Rights Amendment, which enshrines the right to clean air and pure water for all Pennsylvanians, is a guiding principle we should all follow and that is consistent with what RGGI can provide."[40] Time will tell whether the Pennsylvania Supreme Court allows the state to rejoin the RGGI, but the combined forces of various advocacy groups have provided one form of pushback in a complex polycentric system.

To sum up, climate change is not just one problem, it's many interconnected problems at many scales and scopes with many causes and consequences. The knowledge necessary to mitigate the causes and adapt to the consequences is not clear or straightforward, which is why a polycentric approach is critical. If we are to use public policies to address climate change, we must ask ourselves many questions: What kinds of policies? Where should the policies be implemented? What is the correct scale to implement a particular policy? Who has the relevant knowledge? What kinds of incentives will different policies create? What are the potential unintended consequences that might arise? The economic tools presented in this book give us a starting point to answer these questions.

Besides thinking about public policy, the role of markets and civil society can't be understated. Markets are important because they provide many of the new innovations that we need to produce energy more cleanly and efficiently. Civil society groups, like clubs, nonprofits, and other organizations, also provide the support necessary to change the minds of policymakers, business leaders, and consumers. As Elinor Ostrom has argued, "Thus scholars, public officials, and citizens who are concerned with solving collective action problems effectively, equitably, and efficiently, recognize the importance of authorizing citizens to constitute their own local jurisdictions and associations using the knowledge and experience they have about the public problems they face."[41]

Conclusion

Any tool is just a means to an end. We don't use a hammer simply because we think hammering something is good. We use a hammer to build something that we find desirable. In other words, we use tools to accomplish our normative goals. Environmentalists who embrace the "economic tools" discussed in this book can better accomplish their normative goals. They can better understand the root causes behind ecological problems, as well as analyze and evaluate the many possible solutions. Saving endangered species, preserving landscapes, conserving water, and addressing climate change are all real environmental problems, but by using economics as a lens, we can see a path toward a better future.

However, addressing environmental issues is just one important normative consideration, among many. Reasonable people often disagree on how we should prioritize our normative values. We can use the tools of economics to help us think through the compatibilities and potential tensions between environmentalism and other important normative implications, such as facilitating peace, ensuring individual liberty, and promoting democracy.

In terms of facilitating peace, these economic tools show how people can address environmental issues in nonviolent ways when they can engage in positive-sum exchanges. In the real world, people constantly disagree about the seriousness of various environmental problems, and they disagree about the best way to address those problems. Disagreements can become conflict, and conflict can become violent. For example, in recent years, thousands of environmental activists have been the victims of violence while fighting for what they believed in.[1] On the other side, some environmental activists have taken their actions to extremes, using violence to accomplish their goals and even engaging in ecoterrorism.[2]

Avoiding violence when possible is a worthy goal, and people can address pressing environmental problems through peaceful, nonviolent means. When people can engage in cooperative, voluntary, mutually beneficial exchanges, they can come to agreements with one another that avoid the need for violence. When institutions allow positive-sum outcomes to emerge, parties that would otherwise fight can bargain instead. Market exchanges, market-like regulations, and voluntary civil society actions can transform conflict situations into cooperative and collaborative opportunities.

The market process is normatively important because it facilitates nonviolent action on a *micro* and a *macro* scale.[3] At the micro level, individuals in a market choose to be (generally) peaceful because being violent would harm their self-interest. Sellers will not choose to interact with violent buyers and vice versa. At the macro level, the spontaneous order of the market creates a widespread network of nonviolent social relations among diverse people. When many people exchange with one another, they become dependent on one another, making it much costlier for them to fight. In short, a benefit of a market-based spontaneous order is that the combination of many exchanges among people with different (and often opposing) ends produces a broader order of relatively peaceful social relations. When people rely on one another in a market, they create dense networks of mutual interdependence, which means that they have a weaker incentive to fight and a stronger incentive to be peaceful with one another.[4]

Markets aren't perfect, however. Markets can and do have inequality, unbalanced power dynamics, and oppression. The important thing to remember is that different types of institutions are not divided into two simple categories—"violent" and "nonviolent." Instead, there are a wide variety of institutions that have a spectrum of violence. Because the real world is so complex, we need to compare the scale, scope, and types of violence that occur within different institutional contexts. Markets are not perfectly nonviolent, but they are often the least violent means to address conflict over different values.

Besides markets, actions within civil society are important for facilitating peaceful change. Instead of violence, people can engage in persuasion and protest to solve environmental problems, or other controversial social problems.[5] Some of the more common forms are public speaking, picketing, holding vigils, and engaging in mass demonstrations. Boycotts, strikes, civil disobedience, and malicious compliance are also peaceful ways of making nonviolent statements that help persuade decision-makers and other citizens of the need for change.

The alternative to the voluntary actions of markets and civil society is government control, which is often zero-sum and can devolve into negative-sum outcomes. Unless government policies are crafted carefully, the zero-sum nature of a policy means that one side's win is another side's loss. Nobody wants to be the loser, so the losing side fights back. As the stakes become higher for losing, conflict becomes more likely, and in the most extreme circumstances, violence erupts. In the worst forms, scholars have found that political corruption leads to permissive attitudes toward violence, and government corruption is linked to violent conflict through social unrest, social stratification, and a lack of legitimacy in public institutions.[6]

Thus, a polycentric approach limits potential violent conflict because decentralized decision-making in governments, markets, and civil society allows for peaceful, mutually beneficial arrangements to emerge from the complex interactions of many individuals. Different people can choose to cooperate when they face problems that they have in common, but they can also take different approaches when they disagree on means or ends. Values and beliefs vary widely from community to community, but polycentricity allows governance institutions to conform to local values and beliefs, meaning that individuals are more inclined to uphold those institutions. This flexibility lowers the stakes of collective decision-making, reducing the need for conflict and, potentially, violence.

In addition to peace, individual liberty, both political and economic, is a precious thing that many people in modern society take for granted. As economist Peter Boettke defines it, the philosophical idea of liberalism—meaning a focus on individual liberty, not the modern American left-leaning ideologies—is "a doctrine of economic and political life grounded in the recognition that we are one another's dignified equals, and that justice demands equal treatment of equals. No exceptions, no excuses."[7] As economist Deirdre McCloskey puts it,

> Liberalism is liberty from physical coercion by other humans, and in particular a liberty from coercion by masters or governments or gangsters, or masterful governmental gangsters. [...] But yes, I admit it: some imposition by governmental coercion is necessary. Not all laws are bad. Not all taxes, either. Got it. But perhaps then we can move to the question of exactly *how much* law, *how much* taxation, *how much* coercion.[8]

Political and economic liberty are necessary for solving large-scale environmental problems. Many of the examples discussed in this book are the result of people using their liberty to creatively and entrepreneurially discover solutions to environmental problems that they encounter. When individuals in South Africa had the liberty to own and sell rhinos, they were able to bring them back from the brink of extinction. When people in Montana were free to buy and sell willingly, they were able to stitch together the American Prairie Reserve, one of the most ambitious examples of conservation in years. When people in Australia were free to trade their water rights with one another, they found more effective and efficient ways to use the scarce water in their arid landscape.

Relying on top-down, command-and-control policies may sometimes solve environmental problems, but we must ask ourselves, "At what cost?" In other words, sometimes public policies force us to trade off our liberty, which may not be worth it. Of course, we legally restrict liberty all the time—civilians don't have the liberty to own nuclear weapons, and people don't have the liberty to simply take things that belong to others. However, restricting liberty too much has both moral and pragmatic downsides. Morally, stripping away too much liberty is dehumanizing or paternalistic. Pragmatically, restricting individual liberty necessarily limits opportunities for innovation, creativity, and entrepreneurship, which can be used to find potential solutions to environmental problems. Relatedly, protecting people's individual liberties through the rule of law allows people to express their own preferences and use their unique, context-dependent knowledge in addressing the problems they face.

In addition to peace and liberty, democracy is a core normative concern when it comes to addressing ecological issues. Democracy is more than just going to the ballot box or the polling booth every few years. Democracy is really about self-governance—a system in which people come together to form rules for themselves and then monitor and enforce those rules. As political scientist Vincent Ostrom describes it, "Person-to-person, citizen-to-citizen relationships are what life in democratic societies is all about. Democratic ways of life turn into self-organizing and self-governing capabilities rather than presuming that something called 'the Government' governs."[9] He continues, "[E]ach individual is first one's own governor and then capable of working out appropriate patterns of rule ordered associations with others through mutual agreements based on common knowledge and shared understandings as the foundations for conjoint activities and communities of relationships."[10]

What he means is that citizens must be actively involved in governing themselves, and they must participate in decision-making in all the different types of communities to which they belong. Democracy is far more involved than just electing leaders. Instead, it's about collaboratively working together to address shared problems by creating rules and finding effective ways of monitoring and enforcing those rules. This must happen at all the different scales in which an individual is a community member, such as a neighborhood, club, church, school district, city, state, or nation. At every level, effective problem-solving and governance require the direct involvement of the people who are being governed.

When a group of people, large or small, have the freedom to create their own policies and approaches to address their environmental problems, they foster a sense of legitimacy. When individuals feel that "outsiders" are forcing an unwanted policy on them, they may ignore, subvert, or resist the policy. If feelings of resentment against outsiders imposing their will are strong, local communities may turn to violent means of resistance. Polycentric systems involve an element of decentralization, which fosters a sense of legitimacy toward governance because individuals feel as if they (or their authorized representatives) are making the rules, instead of outsiders imposing decisions on them.

In summary, thinking about environmental problems requires a constant dance between positive and normative analysis. You may have deeply rooted moral and ethical convictions about ecological issues—saving species from extinction, preserving untouched landscapes, ensuring that scarce water isn't wasted, and fighting against human-caused climate change. With these normative considerations, the next step is to pivot to the cold, hard analysis of economics, using the tools discussed in this book. Concepts such as incentives, constraints, tradeoffs, and unintended consequences, among many others, help us see why ecological problems arise and what can be done to address them. After we've used economic tools to analyze an environmental problem or attempted solution, we can pivot back to our normative framework to evaluate the outcomes. If the outcomes undermine peace, liberty, and democracy, perhaps we ought to reconsider the approaches we use. The number of approaches to solve ecological problems is limited only by each of our imaginations, so let's use the best of human creativity to find win-win solutions that benefit both people and the environment.

NOTES

Introduction

1. Ian Smith, "Extreme Climate Protesters Made Headlines in 2022. Here's Where They Are Now," *Euro News*, December 28, 2022, https://www.euronews.com/green/2022/12/28/extreme-climate-protesters-made-headlines-in-2022-heres-where-they-are-now.

2. Frank Jordans, "Climate Activists Gluing Themselves to Berlin's Roads Say They Want to Cause 'Peaceful Friction,'" *Euro News*, July 15, 2022, https://www.euronews.com/green/2022/07/15/climate-activists-sticking-themselves-to-berlins-roads-say-they-want-to-cause-peaceful-fri.

3. Smith, "Extreme Climate Protesters Made Headlines in 2022. Here's Where They Are Now."

4. Rachel Ramirez and Ella Nilsen, "Tired but Determined, 5 Young Activists Continue Their Hunger Strike Outside the White House," *CNN*, October 26, 2021, https://www.cnn.com/2021/10/26/us/climate-hunger-strike-white-house/index.html.

5. Ramirez and Nilsen.

6. Gallup, "Environment," 2024, https://news.gallup.com/poll/1615/environment.aspx.

7. Roger E. Backhouse and Steven G. Medema, "Retrospectives: On the Definition of Economics," *The Journal of Economic Perspectives* 23, no. 1 (2009): 221–34.

8. John Maynard Keynes, Elizabeth Johnson, and Donald Moggridge, eds., "Editor and Referee," in *The Collected Writings of John Maynard Keynes: Volume 12: Economic Articles and Correspondence: Investment and Editorial*, The Collected Writings of John Maynard Keynes (London: Royal Economic Society, 1978), 856.

9. Lionel Robbins Robbins, *An Essay on the Nature and Significance of Economic Science*, 2nd ed., rev, and extended (London: Macmillan, 1935).

10. James M. Buchanan, "What Should Economists Do?," in *The Logical Foundations of Constitutional Liberty*, vol. 1, The Collected Works of James M. Buchanan (Indianapolis, IN: Liberty Fund, 1999), 40.

11. Friedrich A. Hayek, *Law, Legislation and Liberty, Volume 2: The Mirage of Social Justice* (Chicago, IL: University of Chicago Press, 1976), 108–9.

12 David M. Levy and Sandra J. Peart, "The Secret History of the Dismal Science. Part I. Economics, Religion and Race in the 19th Century" (The Library of Economics and Liberty, January 22, 2001), https://www.econlib.org/library/Columns/LevyPeartdismal.html.

13 Terry L. Anderson and Donald R. Leal, *Free Market Environmentalism* (San Francisco: Pacific Research Institute for Public Policy, 1991); Terry L. Anderson and R. Donald Leal, *Free Market Environmentalism for the Next Generation*, 1st ed. (New York, NY: Palgrave Macmillan, 2015).

14 Anderson and Leal, *Free Market Environmentalism for the Next Generation*, 6.

15 Anderson and Leal, *Free Market Environmentalism for the Next Generation*, 8.

Chapter 1

1 F. A. Hayek, "The Facts of the Social Sciences," *Ethics* 54, no. 1 (1943): 1–13.

2 Geoffrey Brennan and Loren E. Lomasky, *Democracy and Decision: The Pure Theory of Electoral Preference* (Cambridge, UK: Cambridge University Press, 1993).

3 Peter T. Leeson, *WTF?!: An Economic Tour of the Weird* (Stanford, CA: Stanford Economics and Finance, an imprint of Stanford University Press, 2017); Peter T. Leeson, *The Invisible Hook: The Hidden Economics of Pirates* (Princeton, NJ: Princeton University Press, 2009).

4 Russell Roberts, "Incentives Matter" (The Library of Economics and Liberty, June 5, 2006), https://www.econlib.org/library/Columns/y2006/Robertsincentives.html.

5 Louis Anslow, "In 1903, New York Times Predicted That Airplanes Would Take 10 Million Years to Develop," *Big Think*, October 18, 2023, https://bigthink.com/pessimists-archive/air-space-flight-impossible/.

Chapter 2

1 Oliver Wainwright, "How Solar Farms Took over the California Desert: 'An Oasis Has Become a Dead Sea,'" *The Guardian*, May 21, 2023, https://www.theguardian.com/us-news/2023/may/21/solar-farms-energy-power-california-mojave-desert.

2 Timo Herberz, Claire Y. Barlow, and Matthias Finkbeiner, "Sustainability Assessment of a Single-Use Plastics Ban," *Sustainability* 12, no. 9 (2020).

3   C. S. Oaya et al., "Impact of Synthetic Pesticides Utilization on Humans and the Environment: An Overview," *Agricultural Science and Technology* 11, no. 4 (2019): 279–86; I. Rashmi et al., "Organic and Inorganic Fertilizer Contaminants in Agriculture: Impact on Soil and Water Resources," in *Contaminants in Agriculture: Sources, Impacts and Management*, ed. M. Naeem, Abid Ali Ansari, and Sarvajeet Singh Gill (Cham: Springer International Publishing, 2020), 3–41.

4   Olivia M. Smith et al., "Organic Farming Provides Reliable Environmental Benefits but Increases Variability in Crop Yields: A Global Meta-Analysis," *Frontiers in Sustainable Food Systems* 3 (2019).

Chapter 3

1   Emily L. Tucker et al., "Economic and Life Cycle Assessment of Recycling Municipal Glass as a Pozzolan in Portland Cement Concrete Production," *Resources, Conservation and Recycling* 129 (2018): 240–47; Xuehong Yuan et al., "Integrated Assessment of Economic Benefits and Environmental Impact in Waste Glass Closed-Loop Recycling for Promoting Glass Circularity," *Journal of Cleaner Production* 444 (2024): 141155.

Chapter 4

1   Douglass C. North, *Institutions, Institutional Change, and Economic Performance* (New York City, NY: Cambridge University Press, 1990); Elinor Ostrom, *Understanding Institutional Diversity* (Princeton, NJ: Princeton University Press, 2005).

2   Peter T. Leeson, *The Invisible Hook: The Hidden Economics of Pirates* (Princeton, NJ: Princeton University Press, 2009).

3   Laura E. Grube and Virgil Henry Storr, eds., *Culture and Economic Action*, New Thinking in Political Economy (Cheltenham, UK: Edward Elgar Publishing, 2015).

4   Clifford Geertz, "Thick Description: Toward an Interpretive Theory of Culture," in *The Interpretation of Cultures* (New York, NY: Basic Books, 1973), 3–30.

5   Peter J. Boettke, Christopher J. Coyne, and Peter T. Leeson, "Institutional Stickiness and the New Development Economics," *American Journal of Economics and Sociology* 67, no. 2 (2008): 331–58.

6   Eric Brignac, "The Commerce Clause Justification of Federal Endangered Species Protection: Gibbs v. Babbit," *North Carolina Law Review* 79, no. 3 (2001): 873–87; Justin Gregory Reden, "The Commerce Clause Appropriately Defined Within a University Without Distinction: The Federal Endangered Species Act's Unconstitutional Application to Instrastate Species," *Thomas Jefferson Law Review* 25, no. 3 (Summer 2003): 649–76.

7   Daeun Chloe Shin and Byoungho Ellie Jin, "Do Fur Coats Symbolize Status or Stigma? Examining the Effect of Perceived Stigma on Female Consumers' Purchase Intentions toward Fur Coats," *Fashion and Textiles* 8, no. 1 (February 25, 2021): 10.

Chapter 5

1   National Park Service, "Record of Decision: Bison Management Plan July 2024," July 2024, https://parkplanning.nps.gov/document.cfm?documentID=138363#:~:text=The%20Record%20of%20Decision%2C%20a,previous%20EIS%2C%20completed%20in%202000.

2   Turner Enterprises, Inc, "Flying D Ranch," 2024, https://www.tedturner.com/turner-ranches/turner-ranch-map/flying-d-ranch-montana/.

3   Eriko Hoshino et al., "Individual Transferable Quotas in Achieving Multiple Objectives of Fisheries Management," *Marine Policy* 113 (March 1, 2020): 103744; Cindy Chu, "Thirty Years Later: The Global Growth of ITQs and Their Influence on Stock Status in Marine Fisheries," *Fish and Fisheries (Oxford, England)* 10, no. 2 (2009): 217–30.

4   Samson Afewerki et al., "Innovation in the Norwegian Aquaculture Industry," *Reviews in Aquaculture* 15, no. 2 (2023): 759–71; Jingjie Chu et al., "Stakeholders' Perceptions of Aquaculture and Implications for Its Future: A Comparison of the U.S.A. and Norway," *Marine Resource Economics* 25, no. 1 (2010): 61–76.

5   Elinor Ostrom, *Governing the Commons: The Evolution of Institutions for Collective Action* (New York City, NY: Cambridge University Press, 1990); Elinor Ostrom, "Beyond Markets and States: Polycentric Governance of Complex Economic Systems," *American Economic Review* 100, no. 3 (2010): 641–72.

6   Ronald H. Coase, "The Problem of Social Cost," *The Journal of Law & Economics* 3 (1960): 1–44.

7   Deirdre McCloskey, "The So-Called Coase Theorem," *Eastern Economic Journal* 24, no. 3 (1998): 367–71.

Chapter 6

1. Friedrich A. Hayek, "The Use of Knowledge in Society," *The American Economic Review* 35, no. 4 (1945): 519–30.

2. Don Lavoie, "The Market as a Procedure for Discovery and Conveyance of Inarticulate Knowledge," *Comparative Economic Studies* 28, no. 1 (1986): 1–19.

3. Frank Hyneman Knight, *Risk, Uncertainty and Profit*, Midway Reprint ed. (Chicago, IL: University of Chicago Press, 1985).

4. Hayek, "The Use of Knowledge in Society," 526.

5. Israel M. Kirzner, *Competition and Entrepreneurship* (Chicago, IL: University of Chicago Press, 1973); Israel M. Kirzner, "Entrepreneurial Discovery and the Competitive Market Process: An Austrian Approach," *Journal of Economic Literature* 35, no. 1 (1997): 60–85.

6. Joseph A. Schumpeter, *Capitalism, Socialism and Democracy*, 3rd ed. (New York City, NY: Harper Perennial Modern Thought, 2008).

7. Gordon E. Shockley et al., "Toward a Theory of Public Sector Entrepreneurship," *International Journal of Entrepreneurship and Innovation Management* 6, no. 3 (2006): 205–23.

8. Sarah H. Alvord, L. David Brown, and Christine W. Letts, "Social Entrepreneurship and Societal Transformation: An Exploratory Study," *The Journal of Applied Behavioral Science* 40, no. 3 (2004): 260–82; J. Gregory Dees, Peter Economy, and Jed Emerson, *Enterprising Nonprofits: A Toolkit for Social Entrepreneurs*, Wiley Nonprofit Law, Finance, and Management Series (New York: Wiley, 2001).

9. Virgil Henry Storr, Stefanie Haeffele-Balch, and Laura E. Grube, *Community Revival in the Wake of Disaster: Lessons in Local Entrepreneurship* (New York City, NY: Palgrave Macmillan, 2015).

10. Don Lavoie, *Rivalry and Central Planning* (Arlington, VA: Mercatus Center at George Mason University, 2015); Don Lavoie, *National Economic Planning: What Is Left?* (Arlington, VA: Mercatus Center at George Mason University, 2016); Peter J. Boettke, *The Political Economy of Soviet Socialism: The Formative Years, 1918–1928* (Boston: Kluwer Academic, 1990); Peter J. Boettke and Rosolino A. Candela, "Lessons on Economics and Political Economy from the Soviet Tragedy," *Journal of Global Initiatives: Policy, Pedagogy, Perspective* 12, no. 1 (2018): 32–47.

11. Lavoie, *Rivalry and Central Planning*, 22.

12. Lavoie, 24.

13. Yahya Alshamy et al., "The Market Process as Nonviolent Action," *The Independent Review* 28, no. 1 (2023): 16.

Chapter 8

1. Frédéric Bastiat, "What Is Seen and What Is Not Seen, or Political Economy in One Lesson" (The Online Library of Liberty, 2015), https://oll.libertyfund.org/pages/wswns.

2. Bastiat.

3. Peter J. Boettke, Christopher J. Coyne, and Abigail R. Hall, "Keep off the Grass: The Economics of Prohibition and U.S. Drug Policy," *Oregon Law Review* 91, no. 4 (2013): 1069–95; Jeffrey A. Miron, *Drug War Crimes: The Consequences of Prohibition* (Oakland, CA: The Independent Institute, 2004); Audrey Redford and Angela K. Dills, "The Political Economy of Drug and Alcohol Regulation during the COVID-19 Pandemic," *Southern Economic Journal* 87, no. 4 (2021): 1175–1209; Chad L. Smith, Gregory Hooks, and Michael Lengefeld, "The War on Drugs in Colombia: The Environment, the Treadmill of Destruction and Risk-Transfer Militarism," *Journal of World-Systems Research* 20, no. 2 (2015): 185–206, https://doi.org/10.5195/jwsr.2014.554.

4. Friedrich A. Hayek, "The Results of Human Action but Not of Human Design (1967)," in *The Collected Works of F. A. Hayek, Volume 15: The Market and Other Orders* (Chicago, IL: University of Chicago Press, 2014), 293–303.

5. Adam Smith, *An Inquiry into the Nature and Causes of the Wealth of Nations*, vol. 1 (Indianapolis, IN: Liberty Fund, 1982), 26–27.

6. Smith, 1:456.

7. Hayek, "The Use of Knowledge in Society," 527.

8. Smith, *An Inquiry into the Nature and Causes of the Wealth of Nations*, 1:22–24.

9. Leonard E. Read, *I, Pencil: My Family Tree* (Irvington-on-Hudson, NY: Foundation for Economic Education, 1958).

10. *How to Make a $1500 Sandwich in Only 6 Months*, 2015, https://www.youtube.com/watch?v=URvWSsAgtJE.

Chapter 9

1. James M. Buchanan, "Politics without Romance: A Sketch of Positive Public Choice Theory and Its Normative Implications," in *The Logical Foundations of Constitutional Liberty*, vol. 1, The Collected Works of James M. Buchanan (Indianapolis, IN: Liberty Fund, 1999), 45–59.

2. Brennan and Lomasky, *Democracy and Decision*.

3   Bryan Caplan, *The Myth of the Rational Voter: Why Democracies Choose Bad Policies* (Princeton, NJ: Princeton University Press, 2007).

4   Caplan.

5   Matthew D. Mitchell, *The Pathology of Privilege: The Economic Consequences of Government Favoritism* (Arlington, VA: Mercatus Center at George Mason University, 2014).

6   Randall G. Holcombe, *Political Capitalism: How Economic and Political Power Is Made and Maintained* (Cambridge, UK: Cambridge University Press, 2018); Richard E. Wagner, *Politics as a Peculiar Business* (Cheltenham, UK: Edward Elgar Publishing, 2016).

7   Gordon Tullock, "The Welfare Costs of Tariffs, Monopolies, and Theft," in *Virginia Political Economy*, vol. 1, The Selected Works of Gordon Tullock (Indianapolis, IN: Liberty Fund, 2004), 169–79; Gordon Tullock, "The Transitional Gains Trap," in *Virginia Political Economy*, vol. 1, The Selected Works of Gordon Tullock (Indianapolis, IN: Liberty Fund, 2004).

8   Tullock, "The Transitional Gains Trap"; Tullock, "The Welfare Costs of Tariffs, Monopolies, and Theft"; Mitchell, *The Pathology of Privilege*.

9   Mark A. Groombridge, "America's Bittersweet Sugar Policy" (Center for Trade Policy Studies at the Cato Institute, December 4, 2001), https://cdi.mecon.gob.ar/bases/doc/cato/tbp13.pdf.

10 Adam C. Smith and Bruce Yandle, *Bootleggers and Baptists: How Economics Forces and Moral Persuasion Interact to Shape Regulatory Politics* (Washington, DC: Cate Institute, 2014).

11 Smith and Yandle, 173.

12 James Brooks, "Fourteen Years on, Palin's 'Bridge to Nowhere' Comment Still Resounds," *KTOO*, September 16, 2022, https://www.ktoo.org/2022/09/16/palin-bridge-to-nowhere-gravina-island/; Stephen Dinan, "Alaska Kills 'Bridge to Nowhere' That Helped Put End to Earmarks," *The Washington Times*, November 8, 2015, https://www.washingtontimes.com/news/2015/nov/8/alaska-kills-bridge-to-nowhere-that-helped-put-end/.

13 William A. Niskanen, *Bureaucracy and Representative Government* (New York City, NY: Routledge, 1971).

14 Robert Higgs, *Crisis and Leviathan: Critical Episodes in the Growth of American Government*, 25th anniversary edition (Oakland, CA: Independent Institute, 2012).

15 Gordon Tullock, *Bureaucracy*, vol. 6, The Selected Works of Gordon Tullock (Indianapolis, IN: Liberty Fund, 2005).

Chapter 10

1. Vincent Ostrom, *The Intellectual Crisis in American Public Administration* (Tuscaloosa: University of Alabama Press, 1973); Ostrom, *Governing the Commons*; Vincent Ostrom, *The Meaning of Democracy and the Vulnerability of Democracies: A Response to Tocqueville's Challenge* (Ann Arbor: University of Michigan Press, 1997); Vlad Tarko, *Elinor Ostrom: An Intellectual Biography* (New York City, NY: Rowman and Littlefield, 2017).

2. Jordan K. Lofthouse and Roberta Q. Herzberg, "The Continuing Case for a Polycentric Approach for Coping with Climate Change," *Sustainability (Basel, Switzerland)* 15, no. 4 (2023): 3770.

3. Vincent Ostrom, Charles M. Tiebout, and Robert Warren, "The Organization of Government in Metropolitan Areas: A Theoretical Inquiry," *The American Political Science Review* 55, no. 4 (1961): 831–42.

4. Stephen W. Wang and Maxwell K. Hsu, "Airline Co-Branded Credit Cards—An Application of the Theory of Planned Behavior," *Journal of Air Transport Management* 55 (August 1, 2016): 245–54.

5. Louise C. Druedahl, Timo Minssen, and W. Nicholson Price, "Collaboration in Times of Crisis: A Study on COVID-19 Vaccine R&D Partnerships," *Vaccine* 39, no. 42 (October 8, 2021): 6291–95.

6. James M. Olson, "Navigating the Great Lakes Compact: Water, Public Trust, and International Trade Agreements," *Michigan State Law Review* (2006): 1103–40.

7. Norris Hundley Jr., *Water and the West: The Colorado River Compact and the Politics of Water in the American West*, 2nd ed. (Berkeley, CA: University of California Press, 2009).

8. Dermot Hodson et al., *The Institutions of the European Union*, 5th ed. (Oxford, UK: Oxford University Press, 2022).

9. Elinor Ostrom, "Crossing the Great Divide: Coproduction, Synergy, and Development," *World Development* 24, no. 6 (1996): 1073–87.

10. Paul Dragos Aligica, Peter J. Boettke, and Vlad Tarko, *Public Governance and the Classical-Liberal Perspective* (Oxford, UK: Oxford University Press, 2019).

11. Aligica, Boettke, and Tarko.

12. Stefanie Haeffele and Virgil Henry Storr, *Bottom-Up Responses to Crisis*, Mercatus Studies in Political and Social Economy (Cham, Switzerland: Palgrave Macmillan, 2020).

13. Elinor Ostrom, "Nested Externalities and Polycentric Institutions: Must We Wait for Global Solutions to Climate Change before Taking Actions at Other Scales?," *Economic Theory* 49, no. 2 (2012): 353–69.

Chapter 11

1. Terry L. Anderson and Donald R. Leal, *Free Market Environmentalism for the Next Generation*, 1st ed. (New York, NY: Palgrave Macmillan, 2015), 7.

2. Adam C. Smith and Bruce Yandle, *Bootleggers and Baptists: How Economics Forces and Moral Persuasion Interact to Shape Regulatory Politics* (Washington, DC: Cate Institute, 2014), 109.

3. Smith and Yandle, 112–13.

4. Smith and Yandle, 117–28.

5. Smith and Yandle, 128.

6. Peter Jacobsen and Louis Rouanet, "Economists versus Engineers: Two Approaches to Environmental Problems," *The Review of Austrian Economics* 35, no. 3 (September 1, 2022): 360.

7. Anderson and Leal, *Free Market Environmentalism for the Next Generation*, 3.

8. Anderson and Leal, 31.

9. Anderson and Leal, 42.

10. Anderson and Leal, 163.

11. Anderson and Leal, 163.

12. Elinor Ostrom, "Beyond Markets and States: Polycentric Governance of Complex Economic Systems," *American Economic Review* 100, no. 3 (2010): 664.

13. Anderson and Leal, *Free Market Environmentalism for the Next Generation*, 43–44.

14. Byron Swift, "Command Without Control: Why Cap-and-Trade Should Replace Rate Standards for Regional Pollutants," *Environmental Law Reporter* 31, no. 3 (2001): 10330.

15. A. Denny Ellerman et al., *Markets for Clean Air: The U.S. Acid Rain Program*, 2000, https://doi.org/10.1017/CBO9780511528576.

16. Brian C. Murray and Peter T. Maniloff, "Why Have Greenhouse Emissions in RGGI States Declined? An Econometric Attribution to Economic, Energy Market, and Policy Factors," *Energy Economics* 51 (September 1, 2015): 581–89; Patrick Bayer and Michaël Aklin, "The European Union Emissions Trading System Reduced CO_2 Emissions despite Low Prices," *Proceedings of the National Academy of Sciences—PNAS* 117, no. 16 (2020): 8804–12.

17. Sugandha D. Tuladhar, Sebastian Mankowski, and Paul Bernstein, "Interaction Effects of Market-Based and Command-and-Control Policies," *The Energy Journal (Cambridge, Mass.)* 35, no. 1_suppl (2014): 84.

18 Ann E. Carlson, "Designing Effective Climate Policy: Cap-and-Trade and Complementary Policies," *Harvard Journal on Legislation* 49, no. 2 (2012): 210.

19 Carlson, 221.

20 Anderson and Leal, *Free Market Environmentalism for the next Generation*, 5–6.

21 Robert C. Ellickson, *Order without Law: How Neighbors Settle Disputes* (Cambridge, MA: Harvard University Press, 1991), 4.

22 Anderson and Leal, *Free Market Environmentalism for the Next Generation*, 126.

23 Ostrom, "Beyond Markets and States," 665.

24 Pierre Desrochers and Joanna Szurmak, "The Environmental Benefits of Long-Distance Trade: Insights from the History of By-Product Development," in *The Environmental Optimism of Elinor Ostrom*, ed. Megan E. Jenkins, Randy T. Simmons, and Camille H. Wardle (Logan, UT: Center for Growth and Opportunity at Utah State University, 2020), 173–208.

25 Ellickson, *Order without Law*, 281–82.

26 Anderson and Leal, *Free Market Environmentalism for the Next Generation*, 150.

27 Anderson and Leal, 159.

28 Jess Phelps, "Understanding the Roles of Government in Conservation Easement Transactions," *Denver Law Review* 100, no. 3 (2023): 721–72.

29 Anderson and Leal, *Free Market Environmentalism for the Next Generation*, 142.

30 Virgil Henry Storr, Stefanie Haeffele-Balch, and Laura E. Grube, *Community Revival in the Wake of Disaster: Lessons in Local Entrepreneurship* (New York City, NY: Palgrave Macmillan, 2015), 28.

31 Storr, Haeffele-Balch, and Grube, 32.

32 Storr, Haeffele-Balch, and Grube, 28.

33 Anderson and Leal, *Free Market Environmentalism for the Next Generation*, 45.

34 Pierre Desrochers and Joanna Szurmak, "Population Growth and the Governance of Complex Institutions: People Are More than Mouths to Feed," in *The Environmental Optimism of Elinor Ostrom*, ed. Megan E. Jenkins, Randy T. Simmons, and Camille H. Wardle (Logan, UT: Center for Growth and Opportunity at Utah State University, 2020), 116.

35 Desrochers and Szurmak, 95.

36 Desrochers and Szurmak, 100–101.

37 Vincent Geloso, "Statogenic Climate Change? Julian Simon and Institutions," *The Review of Austrian Economics* 35, no. 3 (September 1, 2022): 344.

38 Desrochers and Szurmak, "Population Growth and the Governance of Complex Institutions," 103.

39 Paul Dragos Aligica and Robert Gabriel Ciobanu, "Julian Simon, the Problem of Socio-Ecological Resilience and the 'Ultimate Resource': A Reinterpretation," *The Review of Austrian Economics* 35, no. 3 (September 1, 2022): 284–85.

40 Peter J. Boettke and Christopher J. Coyne, "The Economic Logic behind the Ultimate Resource," *The Review of Austrian Economics* 35, no. 3 (September 1, 2022): 304.

41 Geloso, "Statogenic Climate Change?," 344.

42 Julian Lincoln Simon, *The Ultimate Resource*, 1st ed. (Princeton, NJ: Princeton University Press, 1981).

43 Geloso, "Statogenic Climate Change?," 344.

44 Gale Pooley and Marian Tupy, "Luck or Insight? The Simon–Ehrlich Bet Re-examined," *Economic Affairs* 40, no. 2 (2020): 278.

45 Pooley and Tupy, 278.

46 Pooley and Tupy, 280.

47 Desrochers and Szurmak, "Population Growth and the Governance of Complex Institutions," 124–25.

48 Aligica and Ciobanu, "Julian Simon, the Problem of Socio-Ecological Resilience and the 'Ultimate Resource'," 289.

49 Jacobsen and Rouanet, "Economists versus Engineers," 360.

50 Boettke and Coyne, "The Economic Logic behind the Ultimate Resource," 304.

51 Jacobsen and Rouanet, "Economists versus Engineers," 361.

52 Geloso, "Statogenic Climate Change?," 347.

53 North, *Institutions, Institutional Change, and Economic Performance*; Deirdre N. McCloskey, *Bourgeois Dignity: Why Economics Can't Explain the Modern World* (Chicago, IL: University of Chicago Press, 2010); Joel Mokyr, *A Culture of Growth : The Origins of the Modern Economy*, The Graz Schumpeter Lectures (Princeton, NJ: Princeton University Press, 2016).

54 Geloso, "Statogenic Climate Change?," 355–56.

55 Geloso, 355–56.

56 Gregory Hooks and Chad L. Smith, "Treadmills of Production and Destruction: Threats to the Environment Posed by Militarism," *Organization & Environment* 18, no. 1 (2005): 19–37; Seyfettin Erdogan et al., "Does Military Expenditure

Impact Environmental Sustainability in Developed Mediterranean Countries?," *Environmental Science and Pollution Research International* 29, no. 21 (2022): 31612–30; Mohammad Ali Rajaeifar et al., "Decarbonize the Military—Mandate Emissions Reporting," *Nature (London)* 611, no. 7934 (2022): 29–32; Chad L. Smith and Michael R. Lengefeld, "The Environmental Consequences of Asymmetric War: A Panel Study of Militarism and Carbon Emissions, 2000–2010," *Armed Forces and Society* 46, no. 2 (2020): 214–37; Perekunah B. Eregha, Xuan Vinh Vo, and Solomon Prince Nathaniel, "Military Spending, Financial Development, and Ecological Footprint in a Developing Country: Insights from Bootstrap Causality and Maki Cointegration," *Environmental Science and Pollution Research International* 29, no. 55 (2022): 83945–55; Unbreen Qayyum, Sohail Anjum, and Samina Sabir, "Armed Conflict, Militarization and Ecological Footprint: Empirical Evidence from South Asia," *Journal of Cleaner Production* 281 (2021): 125299.

Chapter 12

1. Gerardo Ceballos et al., "Accelerated Modern Human–Induced Species Losses: Entering the Sixth Mass Extinction," *Science Advances* 1, no. 5 (n.d.): e1400253; Robert H. Cowie, Philippe Bouchet, and Benoît Fontaine, "The Sixth Mass Extinction: Fact, Fiction or Speculation?," *Biological Reviews of the Cambridge Philosophical Society* 97, no. 2 (2022): 640–63.

2. US Fish and Wildlife Service, "Critical Habitat: What Is It?," March 2017, https://www.fws.gov/sites/default/files/documents/critical-habitat-fact-sheet.pdf.

3. Dean Lueck and Jeffrey A. Michael, "Preemptive Habitat Destruction under the Endangered Species Act," *The Journal of Law & Economics* 46, no. 1 (2003): 29.

4. Robert Innes, "The Economics of Takings and Compensation When Land and Its Public Use Value Are in Private Hands," *Land Economics* 76, no. 2 (2000): 195–212; Gardner M. Brown and Jason F. Shogren, "Economics of the Endangered Species Act," *The Journal of Economic Perspectives* 12, no. 3 (1998): 3–20.

5. US Fish and Wildlife Service, "Safe Harbor Agreements," 2024, https://www.fws.gov/service/safe-harbor-agreements; US Fish and Wildlife Service, "Conservation Benefit Agreements," 2024, https://www.fws.gov/service/conservation-benefit-agreements.

6. US Fish and Wildlife Service, "Habitat Conservation Plans," 2024, https://www.fws.gov/service/habitat-conservation-plans.

7. US Fish and Wildlife Service, "Recovery Land Acquisition Grants," 2024, https://www.fws.gov/service/recovery-land-acquisition-grants.

8   Martin F. J. Taylor, Kieran F. Suckling, and Jeffrey J. Rachlinski, "The Effectiveness of the Endangered Species Act: A Quantitative Analysis," *Bioscience* 55, no. 4 (2005): 360–67; J. Michael Scott et al., "Recovery of Imperiled Species under the Endangered Species Act: The Need for a New Approach," *Frontiers in Ecology and the Environment* 3, no. 7 (2005): 383–89; Katherine E. Gibbs and David J. Currie, "Protecting Endangered Species: Do the Main Legislative Tools Work?," *PloS One* 7, no. 5 (2012): e35730.

9   Paul Mackun, Joshua Comenetz, and Lindsay Spell, "Around Four-Fifths of All U.S. Metro Areas Grew Between 2010 and 2020" (United States Census Bureau, August 12, 2021), https://www.census.gov/library/stories/2021/08/more-than-half-of-united-states-counties-were-smaller-in-2020-than-in-2010.html#:~:text=State%20 Population%20Change&text=These%20five%20states%20accounted%20for,grew%20 by%2015.0%25%20or%20more.

10 Mary M. Rowland et al., "Greater Sage-Grouse as an Umbrella Species for Sagebrush-Associated Vertebrates," *Biological Conservation* 129, no. 3 (May 1, 2006): 323–35.

11 Jordan K. Lofthouse and Megan E. Jenkins, "Cooperation or Conflict: Two Approaches to Conservation," in *Regulation and Economic Opportunity: Blueprints for Reform* (Logan, UT: Center for Growth and Opportunity at Utah State University, 2020), 247–76; Jordan K. Lofthouse, "Self-Governance, Polycentricity, and Environmental Policy," in *The Environmental Optimism of Elinor Ostrom* (Logan, UT: Center for Growth and Opportunity at Utah State University, 2020), 31–60.

12 Working Lands for Wildlife, "How We Work" (USDA Natural Resources Conservation Service, 2024), https://www.wlfw.org/about-us/how-we-work/.

13 "The Sage Grouse Initiative" (Working Lands For Wildlife, 2024), https://www.wlfw.org/wildlife/sage-grouse/about-the-initiative/; Lofthouse and Jenkins, "Cooperation or Conflict"; Lofthouse, "Self-Governance, Polycentricity, and Environmental Policy."

14 Terry A. Messmer et al., "Utah's Adaptive Resources Management Greater Sage-Grouse Local Working Groups 2018 Annual Report" (Jack H. Berryman Institute, Department of Wildland Resources, and Utah State University Extension, Logan, March 2019), https://extension.usu.edu/utahcbcp/files/2018LWGAnnualReportMarch2019.pdf; Lorien R. Belton, S. Nicole Frey, and David K. Dahlgren, "Participatory Research in Sage-Grouse Local Working Groups: Case Studies from Utah," *Human-Wildlife Interactions* 11, no. 3 (2017): 287–301; Lofthouse and Jenkins, "Cooperation or Conflict"; Lofthouse, "Self-Governance, Polycentricity, and Environmental Policy."

15 Hayley S. Clements, Dave Balfour, and Enrico Di Minin, "Importance of Private and Communal Lands to Sustainable Conservation of Africa's Rhinoceroses," *Frontiers in Ecology and the Environment* 21, no. 3 (2023): 141.

16 Michael 't Sas-Rolfes, "Saving African Rhinos: A Market Success Story" (PERC, n.d.), https://www.perc.org/wp-content/uploads/2011/08/Saving-African-Rhinos-final.pdf.

17 't Sas-Rolfes.

18 't Sas-Rolfes.

19 't Sas-Rolfes.

20 't Sas-Rolfes.

21 Clements, Balfour, and Di Minin, "Importance of Private and Communal Lands to Sustainable Conservation of Africa's Rhinoceroses," 140.

22 Clements, Balfour, and Di Minin, 140–41.

23 't Sas-Rolfes, "Saving African Rhinos: A Market Success Story," 5–6.

24 't Sas-Rolfes, 6–7.

25 Clements, Balfour, and Di Minin, "Importance of Private and Communal Lands to Sustainable Conservation of Africa's Rhinoceroses," 141.

26 Laura A. Chapman and Piran C. L. White, "Stakeholder Perspectives on the Value and Challenges of Private Rhinoceros Ownership in South Africa," *Human Dimensions of Wildlife* 25, no. 2 (2020): 187–97.

27 Clements, Balfour, and Di Minin, "Importance of Private and Communal Lands to Sustainable Conservation of Africa's Rhinoceroses," 142.

28 Clements, Balfour, and Di Minin, 143.

29 Rubino and Pienaar, "Rhinoceros Ownership and Attitudes towards Legalization of Global Horn Trade within South Africa's Private Wildlife Sector," 247.

30 Clements, Balfour, and Di Minin, 143.

31 Clements, Balfour, and Di Minin, 144.

32 Clements, Balfour, and Di Minin, 144.

33 Sam M. Ferreira, Michele Pfab, and Mike Knight, "Management Strategies to Curb Rhino Poaching: Alternative Options Using a Cost-Benefit Approach," *South African Journal of Science* 110, no. 5–6 (2014): 6.

34 Clements, Balfour, and Di Minin, "Importance of Private and Communal Lands to Sustainable Conservation of Africa's Rhinoceroses," 145.

35 Elena C. Rubino and Elizabeth F. Pienaar, "Rhinoceros Ownership and Attitudes towards Legalization of Global Horn Trade within South Africa's Private Wildlife Sector," *Oryx* 54, no. 2 (2020): 248.

36 Chapman and White, "Stakeholder Perspectives on the Value and Challenges of Private Rhinoceros Ownership in South Africa."

37 Rubino and Pienaar, "Rhinoceros Ownership and Attitudes towards Legalization of Global Horn Trade within South Africa's Private Wildlife Sector," 250.

38 Clements, Balfour, and Di Minin, "Importance of Private and Communal Lands to Sustainable Conservation of Africa's Rhinoceroses," 145.

39 Chapman and White, "Stakeholder Perspectives on the Value and Challenges of Private Rhinoceros Ownership in South Africa."

40 Rubino and Pienaar, "Rhinoceros Ownership and Attitudes towards Legalization of Global Horn Trade within South Africa's Private Wildlife Sector," 247.

41 Ferreira, Pfab, and Knight, "Management Strategies to Curb Rhino Poaching," 6.

42 Clements, Balfour, and Di Minin, "Importance of Private and Communal Lands to Sustainable Conservation of Africa's Rhinoceroses," 142.

43 Clements, Balfour, and Di Minin, 145.

Chapter 13

1 National Park Service, "Hidden Gems, off-Season Trips Highlights of 325.5 Million National Park Visits in 2023," February 22, 2024, https://www.nps.gov/orgs/1207/hidden-gems-off-season-trips-highlights-of-325-5-million-national-park-visits-in-2023.htm#:~:text=Today%2C%20the%20National%20Park%20Service,million%20or%204%25%20over%202022.

2 US Fish and Wildlife Service, "Arctic National Wildlife Refuge," 2024, https://www.fws.gov/refuge/arctic/about-us; US Fish and Wildlife Service, "Mille Lacs National Wildlife Refuge," 2024, https://www.fws.gov/refuge/mille-lacs#:~:text=Mille%20Lacs%20National%20Wildlife%20Refuge%20was%20established%20in%201915%20and,Refuge%20System%20at%200.57%20acres.

3 Kalifi Ferretti-Gallon et al., "National Parks Best Practices: Lessons from a Century's Worth of National Parks Management," *International Journal of Geoheritage and Parks* 9, no. 3 (September 1, 2021): 335–46.

4 Ferretti-Gallon et al., 342.

5 CBC Radio, "Canadian Who Put Baby Bison in SUV Defends His Decision after Court Order," *CBC/Radio-Canada*, June 3, 2016.

6 Associated Press, "A Yellowstone Visitor Picked up a Baby Bison. It Was Put to Death after Its Herd Rejected It," *PBS News*, May 24, 2023.

7 Ryan M. Yonk and Jordan K. Lofthouse, "A Review on the Manufacturing of a National Icon: Institutions and Incentives in the Management of Yellowstone National Park," *International Journal of Geoheritage and Parks* 8, no. 2 (2020): 87–95.

8   Sean Kammer, "Coming to Terms with Wilderness: The Wilderness Act and the Problem of Wildlife Restoration," *Environmental Law* 43, no. 1 (2013): 83–124.

9   "American Prairie: Restoring Bison to Northern Montana with a Patchwork Nature Reserve | 60 Minutes," *60 Minutes*, August 14, 2023, https://www.youtube.com/watch?v=PyKnFHcYk7E.

10 "American Prairie: Restoring Bison to Northern Montana with a Patchwork Nature Reserve | 60 Minutes."

11 Jordan K. Lofthouse and Megan E. Jenkins, "Cooperation or Conflict: Two Approaches to Conservation," in *Regulation and Economic Opportunity: Blueprints for Reform* (Logan, UT: Center for Growth and Opportunity at Utah State University, 2020), 247–76.

12 "American Prairie: Restoring Bison to Northern Montana with a Patchwork Nature Reserve | 60 Minutes."

13 *Last Wild Place: American Prairie Reserve | National Geographic*, YouTube Video, 2020, https://www.youtube.com/watch?v=_Q673Utczgs.

14 "American Prairie: Restoring Bison to Northern Montana with a Patchwork Nature Reserve | 60 Minutes."

15 American Prairie Reserve, "Wild Sky," 2024, https://americanprairie.org/project/wild-sky/.

16 *Last Wild Place: American Prairie Reserve | National Geographic*.

17 "American Prairie: Restoring Bison to Northern Montana with a Patchwork Nature Reserve | 60 Minutes."

18 "American Prairie: Restoring Bison to Northern Montana with a Patchwork Nature Reserve | 60 Minutes."

19 *Last Wild Place: American Prairie Reserve | National Geographic*.

20 "American Prairie: Restoring Bison to Northern Montana with a Patchwork Nature Reserve | 60 Minutes."

21 "American Prairie: Restoring Bison to Northern Montana with a Patchwork Nature Reserve | 60 Minutes."

22 "American Prairie: Restoring Bison to Northern Montana with a Patchwork Nature Reserve | 60 Minutes."

23 "American Prairie: Restoring Bison to Northern Montana with a Patchwork Nature Reserve | 60 Minutes."

24 Bureau of Land Management, "BLM Issues Final Decision on Bison Grazing Proposal," July 28, 2022, https://www.blm.gov/press-release/blm-issues-final-decision-bison-grazing-proposal.

25 United Property Owners of Montana, "Stop APR Legal Action Fund," n.d., https://upom.org/.

Chapter 14

1 Terry L. Anderson and Donald R. Leal, *Free Market Environmentalism for the Next Generation*, 1st ed. (New York, NY: Palgrave Macmillan, 2015), 87.

2 Maruge Zhao, Tiho Ancev, and R. Willem Vervoort, "Water Market Functionality: Evidence from the Australian Experience," *Water Resources Research* 60, no. 2 (2024): 1.

3 Sarah Ann Wheeler et al., "Developing a Water Market Readiness Assessment Framework," *Journal of Hydrology* 552 (September 1, 2017): 807.

4 Wheeler et al., 807.

5 Wheeler et al., 808.

6 Wheeler et al., 808.

7 Michael D. Young, "Unbundling Water Rights as a Means to Improve Water Markets in Australia's Southern Connected Murray-Darling Basin," in *Use of Economic Instruments in Water Policy: Insights from International Experience*, Global Issues in Water Policy, vol.14 (Switzerland: Springer International Publishing AG, 2015), 279.

8 Wheeler et al., "Developing a Water Market Readiness Assessment Framework," 809.

9 Young, "Unbundling Water Rights as a Means to Improve Water Markets in Australia's Southern Connected Murray-Darling Basin," 281.

10 Wheeler et al., "Developing a Water Market Readiness Assessment Framework," 809.

11 Zhao, Ancev, and Vervoort, "Water Market Functionality," 3.

12 Michael D. Young and Jim C. McColl, "Double Trouble: The Importance of Accounting for and Defining Water Entitlements Consistent with Hydrological Realities," *The Australian Journal of Agricultural and Resource Economics* 53, no. 1 (2009): 20.

13 Zhao, Ancev, and Vervoort, "Water Market Functionality," 3.

14 Young, "Unbundling Water Rights as a Means to Improve Water Markets in Australia's Southern Connected Murray-Darling Basin," 282.

15 Zhao, Ancev, and Vervoort, "Water Market Functionality," 3.

16 Young, "Unbundling Water Rights as a Means to Improve Water Markets in Australia's Southern Connected Murray-Darling Basin," 282.

17 Young, 282.

18 Young, 286.

19 Zhao, Ancev, and Vervoort, "Water Market Functionality," 3.

20 Zhao, Ancev, and Vervoort, 26.

21 Young, "Unbundling Water Rights as a Means to Improve Water Markets in Australia's Southern Connected Murray-Darling Basin," 293–94.

22 Wheeler et al., "Developing a Water Market Readiness Assessment Framework," 808.

23 Cameron Holley and Darren Sinclair, "Rethinking Australian Water Law and Governance: Successes, Challenges and Future Directions," *Environmental and Planning Law Journal* 33, no. 4 (July 1, 2016): 275–83.

24 Carl J. Bauer, "Water Conflicts and Entrenched Governance Problems in Chile's Market Model," *Water Alternatives* 8, no. 2 (2015): 147–72; Manuel Prieto, "Indigenous Resurgence, Identity Politics, and the Anticommodification of Nature: The Chilean Water Market and the Atacameño People," *Annals of the American Association of Geographers* 112, no. 2 (2022): 487–504.

25 Parth J. Shah and Vidisha Maitra, eds., *Terracotta Reader: A Market Approach to the Environment* (New Delhi, India: Academic Foundation in association with Centre for Civil Society, New Delhi, 2005), 293.

26 Shah and Maitra.

27 Kenneth A. Baerenklau, Kurt A. Schwabe, and Ariel Dinar, "The Residential Water Demand Effect of Increasing Block Rate Water Budgets," *Land Economics* 90, no. 4 (2014): 683–99.

28 Los Angeles Department of Water and Power, "Residential Water Rates," 2024, https://www.ladwp.com/account/understanding-your-rates/residential-water-rates#:~:text=If%20you%20are%20an%20apartment,costly%20sources%20of%20water%20supply.

29 *As Utah's Great Salt Lake Dries Up, Economic Crisis Looms | WSJ*, 2022, https://www.youtube.com/watch?v=k6oSNLkIBEg.

30 "Great Salt Lake Watershed Enhancement Trust," 2024, https://www.gslwatertrust.org/.

31 Shaela Adams, "PRESS RELEASE: Trust Awards Funding to Protect and Restore More Than 13,000 Acres of Essential Great Salt Lake Wetlands" (Great Salt Lake Watershed Enhancement Trust, November 7, 2023), https://www.gslwatertrust.org/news/press-release-trust-awards-funding-to-protect-and-restore-more-than-13000-acres-of-essential-great-salt-lake-wetlandsnbsp.

32 Todd Wilkinson, "To Save a Landmark Lake" (PERC, July 29, 2024), https://perc.org/2024/07/29/to-save-a-landmark-lake/.

33 Tim Hawkes, "Trout Water," *PERC Reports* 41, no. 2 (2023): 38–45.

Chapter 15

1 Katharina van Baal, Stephanie Stiel, and Peter Schulte, "Public Perceptions of Climate Change and Health-A Cross-Sectional Survey Study," *International Journal of Environmental Research and Public Health* 20, no. 2 (2023): 1464.

2 Alec Tyson, Cary Funk, and Brian Kennedy, "What the Data Says about Americans' Views of Climate Change" (Pew Research Center, August 9, 2023), https://www.pewresearch.org/short-reads/2023/08/09/what-the-data-says-about-americans-views-of-climate-change/.

3 Pew Research Center, "Public Views About Science in Japan," September 29, 2020, https://www.pewresearch.org/science/fact-sheet/public-views-about-science-in-japan/.

4 IPCC, *Climate Change 2022: Impacts, Adaptation and Vulnerability* (Cambridge, UK: Cambridge University Press, 2022), https://www.ipcc.ch/report/ar6/wg2/downloads/report/IPCC_AR6_WGII_FullReport.pdf.

5 Maurie J. Cohen, "Is the UK Preparing for 'War'? Military Metaphors, Personal Carbon Allowances, and Consumption Rationing in Historical Perspective," *Climatic Change* 104, no. 2 (2011): 199–222; Laurence L. Delina, *Strategies for Rapid Climate Mitigation: Wartime Mobilisation as a Model for Action?*, Routledge Advances in Climate Change Research (Abingdon, Oxon: Routledge, Earthscan, 2016); Stephen J. Flusberg, Teenie Matlock, and Paul H. Thibodeau, "Metaphors for the War (or Race) against Climate Change," *Environmental Communication* 11, no. 6 (2017): 769–83; Johannes Kester and Benjamin K. Sovacool, "Torn between War and Peace: Critiquing the Use of War to Mobilize Peaceful Climate Action," *Energy Policy* 104 (2017): 50–55; Andreas Malm, *Corona, Climate, Chronic Emergency: War Communism in the Twenty-First Century*, Coronavirus Pamphlet Series (London: Verso, 2020); Michael E. Mann, *The New Climate War: The Fight to Take Back Our Planet*, 1st ed. (New York: PublicAffairs, 2021).

6 Holcombe, *Political Capitalism*; Mitchell, *The Pathology of Privilege*.

7 Ostrom, "Nested Externalities and Polycentric Institutions"; Elinor Ostrom, "A Polycentric Approach for Coping with Climate Change," *Annals of Economics and Finance* 15, no. 1 (2014): 97–134.

8   Ostrom, "A Polycentric Approach for Coping with Climate Change," 123.

9   Daniel H. Cole, "Advantages of a Polycentric Approach to Climate Change Policy," *Nature Climate Change* 5, no. 2 (2015): 114–18; Lofthouse and Herzberg, "The Continuing Case for a Polycentric Approach for Coping with Climate Change"; Elinor Ostrom, "A Multi-Scale Approach to Coping with Climate Change and Other Collective Action Problems," *Leviathan (Düsseldorf)* 39, no. 3 (2011): 447–58; Ostrom, "Nested Externalities and Polycentric Institutions"

10 Timothy Johnson and Ameya Joshi, "Review of Vehicle Engine Efficiency and Emissions," *SAE International Journal of Engines* 11, no. 6 (2018): 1307–30.

11 Sida Feng and Christopher L. Magee, "Technological Development of Key Domains in Electric Vehicles: Improvement Rates, Technology Trajectories and Key Assignees," *Applied Energy* 260 (February 15, 2020): 114264.

12 Boris R. Lukanov and Elena M. Krieger, "Distributed Solar and Environmental Justice: Exploring the Demographic and Socio-Economic Trends of Residential PV Adoption in California," *Energy Policy* 134 (November 1, 2019): 110935.

13 Philipp Beiter et al., "Wind Power Costs Driven by Innovation and Experience with Further Reductions on the Horizon," *Wiley Interdisciplinary Reviews. Energy and Environment* 10, no. 5 (2021): e398.

14 Hadrien Bajolle, Marion Lagadic, and Nicolas Louvet, "The Future of Lithium-Ion Batteries: Exploring Expert Conceptions, Market Trends, and Price Scenarios," *Energy Research & Social Science* 93 (November 1, 2022): 102850.

15 Shahid Ali, Rodney A. Stewart, and Oz Sahin, "Drivers and Barriers to the Deployment of Pumped Hydro Energy Storage Applications: Systematic Literature Review," *Cleaner Engineering and Technology* 5 (December 1, 2021): 100281; Yimeng Huang and Ju Li, "Key Challenges for Grid-Scale Lithium-Ion Battery Energy Storage," *Advanced Energy Materials* 12, no. 48 (2022): 2202197; Catarina R. Matos, Patrícia P. Silva, and Júlio F. Carneiro, "Overview of Compressed Air Energy Storage Projects and Regulatory Framework for Energy Storage," *Journal of Energy Storage* 55 (November 30, 2022): 105862.

16 Helen Stopps and Marianne F. Touchie, "Residential Smart Thermostat Use: An Exploration of Thermostat Programming, Environmental Attitudes, and the Influence of Smart Controls on Energy Savings," *Energy and Buildings* 238 (May 1, 2021): 110834.

17 Chidiebere Emmanuel Eze et al., "Green Building Materials Products and Service Market in the Construction Industry," *Journal of Engineering, Project, and Production Management* 11, no. 2 (2021): 89–101.

18 Emily Duncan et al., "New but for Whom? Discourses of Innovation in Precision Agriculture," *Agriculture and Human Values* 38, no. 4 (2021): 1181–99.

19. Lee Beck, "Carbon Capture and Storage in the USA: The Role of US Innovation Leadership in Climate-Technology Commercialization," *Clean Energy* 4, no. 1 (April 4, 2020): 2–11.

20. *The Climate Crisis: Can Smart Ideas Save the Planet? | DW Documentary* (DW Documentary, 2023), https://www.youtube.com/watch?v=afPVy0yiLRw.

21. *The Climate Crisis: Can Smart Ideas Save the Planet? | DW Documentary*.

22. *The Climate Crisis: Can Smart Ideas Save the Planet? | DW Documentary*.

23. *What Seaweed and Cow Burps Have to Do with Climate Change | Ermias Kebreab | TED Countdown* (TED, 2022), https://www.youtube.com/watch?v=fCpKG1QJC14.

24. Karen A. Beauchemin et al., "Invited Review: Current Enteric Methane Mitigation Options," *Journal of Dairy Science* 105, no. 12 (2022): 9297–9326, https://doi.org/10.3168/jds.2022-22091.

25. Beauchemin et al.

26. Diana C. Reyes et al., "Maine Organic Dairy Producers' Receptiveness to Seaweed Supplementation and Effect of Chondrus Crispus on Enteric Methane Emissions in Lactating Cows," *Frontiers in Veterinary Science* 10 (2023): 1153097.

27. *The Climate Crisis: Can Smart Ideas Save the Planet? | DW Documentary*.

28. California Air Resources Board, "Cap-and-Trade Program: About," 2024, https://ww2.arb.ca.gov/our-work/programs/cap-and-trade-program/about.

29. California Air Resources Board, "Chapter 1: How Does the Cap-and-Trade Program Work?," September 2012, 1, https://ww2.arb.ca.gov/sites/default/files/cap-and-trade/guidance/chapter1.pdf.

30. Christian Lessmann and Niklas Kramer, "The Effect of Cap-and-Trade on Sectoral Emissions: Evidence from California," *Energy Policy* 188 (May 1, 2024): 114066.

31. Danae Hernandez-Cortes and Kyle C. Meng, "Do Environmental Markets Cause Environmental Injustice? Evidence from California's Carbon Market," *Journal of Public Economics* 217 (January 1, 2023): 104786.

32. Hernandez-Cortes and Meng.

33. Hernandez-Cortes and Meng.

34. Jingchi Yan, "The Impact of Climate Policy on Fossil Fuel Consumption: Evidence from the Regional Greenhouse Gas Initiative (RGGI)," *Energy Economics* 100 (August 1, 2021): 105333.

35. Ceres, "The Regional Greenhouse Gas Initiative: A Fact Sheet," 2016, https://assets.ceres.org/sites/default/files/Fact%20Sheets%20or%20misc%20files/RGGI%20Fact%20Sheet.pdf.

36 Yan, "The Impact of Climate Policy on Fossil Fuel Consumption."

37 Murray and Maniloff, "Why Have Greenhouse Emissions in RGGI States Declined?"

38 Maxwell T. Dorman, Aaron L. Strong, and Nicola Ulibarri, "Coalitions in Climate Mitigation Policy Re-Design Processes: The Case of the Regional Greenhouse Gas Initiative," *Environmental Science & Policy* 127 (January 1, 2022): 44.

39 Clean Air Council, "PA Supreme Court Grants Intervention to Environmental Organizations in RGGI Litigation," July 18, 2024, https://cleanair.org/rggi-intervention/; Regional Greenhouse Gas Initiative, "Materials on New Participation," 2024, https://www.rggi.org/program-overview-and-design/new-participation.

40 Amanda Leland, "RGGI Is the Optimal Path for Pennsylvania to Cut Climate Pollution, Protect Consumers and Win Jobs in the New Clean Economy: Statement of EDF Executive Director Amanda Leland" (Environmental Defense Fund, November 1, 2023), https://www.edf.org/media/rggi-optimal-path-pennsylvania-cut-climate-pollution-protect-consumers-and-win-jobs-new-clean.

41 Ostrom, "A Polycentric Approach for Coping with Climate Change," 127.

Conclusion

1 Emily Kwong, Brit Hanson, and Madeline K. Sofia, "A Rising Tide of Violence Against Environmental Activists," *NPR*, April 14, 2021, https://www.npr.org/2021/04/12/986451622/a-rising-tide-of-violence-against-environmental-activists; Ali Hines, "Decade of Defiance" (Global Witness, May 10, 2023), https://www.globalwitness.org/en/campaigns/environmental-activists/decade-defiance/#decade-killings-globally.

2 Jennifer Varriale Carson, Gary LaFree, and Laura Dugan, "Terrorist and Non-Terrorist Criminal Attacks by Radical Environmental and Animal Rights Groups in the United States, 1970-2007," *Terrorism and Political Violence* 24, no. 2 (2012): 295–319; Jennifer Varriale Carson, "Counterterrorism and Radical Eco-Groups: A Context for Exploring the Series Hazard Model," *Journal of Quantitative Criminology* 30, no. 3 (2014): 485–504; Steven M. Chermak et al., "An Overview of Bombing and Arson Attacks by Environmental and Animal Rights Extremists in the United States, 1995–2010" (Final Report to the Resilient Systems Division, Science and Technology Directorate, US Department of Homeland Security, 2013), https://www.dhs.gov/sites/default/files/publications/OPSR_TP_TEVUS_Bombing-Arson-Attacks_Environmental-Animal%20Rights-Extremists_1309-508.pdf.

3 Alshamy et al., "The Market Process as Nonviolent Action."

4 Alshamy et al.

5 Gene Sharp, *The Politics of Nonviolent Action*, Extending Horizons Books (Boston: P. Sargent Publisher, 1973).

6 Shaazka Beyerle, "Civil Resistance and the Corruption-Violence Nexus," *Journal of Sociology and Social Welfare* 38, no. 2 (2011): 53–78; Robert Gillanders and Lisa Werff, "Corruption Experiences and Attitudes to Political, Interpersonal, and Domestic Violence," *Governance (Oxford)* 35, no. 1 (2022): 167–85.

7 Peter J. Boettke, *The Struggle for a Better World* (Arlington, VA: Mercatus Center at George Mason University, 2021), 6.

8 Deirdre Nansen McCloskey, *Why Liberalism Works: How True Liberal Values Produce a Freer, More Equal, Prosperous World for All* (New Haven, CT: Yale University Press, 2019), 25.

9 Ostrom, *The Meaning of Democracy and the Vulnerability of Democracies*, 3–4.

10 Ostrom, 59.

SELECT BIBLIOGRAPHY

Afewerki, Samson, Frank Asche, Bård Misund, Trine Thorvaldsen, and Ragnar Tveteras. 2023. "Innovation in the Norwegian Aquaculture Industry." *Reviews in Aquaculture* 15 (2): 759–71.

Ali, Shahid, Rodney A. Stewart, and Oz Sahin. 2021. "Drivers and Barriers to the Deployment of Pumped Hydro Energy Storage Applications: Systematic Literature Review." *Cleaner Engineering and Technology* 5 (December): 100281.

Aligica, Paul Dragos, Peter J. Boettke, and Vlad Tarko. 2019. *Public Governance and the Classical-Liberal Perspective*. Oxford, UK: Oxford University Press.

Aligica, Paul Dragos, and Robert Gabriel Ciobanu. 2022. "Julian Simon, the Problem of Socio-Ecological Resilience and the 'Ultimate Resource': A Reinterpretation." *The Review of Austrian Economics* 35 (3): 283–301.

Alshamy, Yahya, Joshua Ammons, Christopher J. Coyne, and Nathan P. Goodman. 2023. "The Market Process as Nonviolent Action." *The Independent Review* 28 (1): 11–28.

Alvord, Sarah H., L. David Brown, and Christine W. Letts. 2004. "Social Entrepreneurship and Societal Transformation: An Exploratory Study." *The Journal of Applied Behavioral Science* 40 (3): 260–82.

Anderson, Terry L., and Donald R. Leal. 1991. *Free Market Environmentalism*. San Francisco: Pacific Research Institute for Public Policy.

Anderson, Terry L., and Donald R. Leal. 2015. *Free Market Environmentalism for the Next Generation*. 1st ed. New York, NY: Palgrave Macmillan.

Baal, Katharina van, Stephanie Stiel, and Peter Schulte. 2023. "Public Perceptions of Climate Change and Health-A Cross-Sectional Survey Study." *International Journal of Environmental Research and Public Health* 20 (2): 1464.

Baerenklau, Kenneth A., Kurt A. Schwabe, and Ariel Dinar. 2014. "The Residential Water Demand Effect of Increasing Block Rate Water Budgets." *Land Economics* 90 (4): 683–99.

Bajolle, Hadrien, Marion Lagadic, and Nicolas Louvet. 2022. "The Future of Lithium-Ion Batteries: Exploring Expert Conceptions, Market Trends, and Price Scenarios." *Energy Research & Social Science* 93 (November): 102850.

Bastiat, Frédéric. 2015. "What Is Seen and What Is Not Seen, or Political Economy in One Lesson." The Online Library of Liberty. https://oll.libertyfund.org/pages/wswns.

Bauer, Carl J. 2015. "Water Conflicts and Entrenched Governance Problems in Chile's Market Model." *Water Alternatives* 8 (2): 147–72.

Bayer, Patrick, and Michaël Aklin. 2020. "The European Union Emissions Trading System Reduced CO_2 Emissions despite Low Prices." *Proceedings of the National Academy of Sciences – PNAS* 117 (16): 8804–12.

Beauchemin, Karen A., Emilio M. Ungerfeld, Adibe L. Abdalla, Clementina Alvarez, Claudia Arndt, Philippe Becquet, Chaouki Benchaar, et al. 2022. "Invited Review:

Current Enteric Methane Mitigation Options." *Journal of Dairy Science* 105 (12): 9297–9326.

Beck, Lee. 2020. "Carbon Capture and Storage in the USA: The Role of US Innovation Leadership in Climate-Technology Commercialization." *Clean Energy* 4 (1): 2–11.

Beiter, Philipp, Aubryn Cooperman, Eric Lantz, Tyler Stehly, Matt Shields, Ryan Wiser, Thomas Telsnig, Lena Kitzing, Volker Berkhout, and Yuka Kikuchi. 2021. "Wind Power Costs Driven by Innovation and Experience with Further Reductions on the Horizon." *Wiley Interdisciplinary Reviews. Energy and Environment* 10 (5): e398.

Belton, Lorien R., S. Nicole Frey, and David K. Dahlgren. 2017. "Participatory Research in Sage-Grouse Local Working Groups: Case Studies from Utah." *Human-Wildlife Interactions* 11 (3): 287–301.

Beyerle, Shaazka. 2011. "Civil Resistance and the Corruption-Violence Nexus." *Journal of Sociology and Social Welfare* 38 (2): 53–78.

Boettke, Peter J. 1990. *The Political Economy of Soviet Socialism: The Formative Years, 1918-1928*. Boston: Kluwer Academic.

Boettke, Peter J. 2021. *The Struggle for a Better World*. Arlington, VA: Mercatus Center at George Mason University.

Boettke, Peter J., and Rosolino A. Candela. 2018. "Lessons on Economics and Political Economy from the Soviet Tragedy." *Journal of Global Initiatives: Policy, Pedagogy, Perspective* 12 (1): 32–47.

Boettke, Peter J., and Christopher J. Coyne. 2022. "The Economic Logic behind the Ultimate Resource." *The Review of Austrian Economics* 35 (3): 303–14.

Boettke, Peter J., Christopher J. Coyne, and Abigail R. Hall. 2013. "Keep off the Grass: The Economics of Prohibition and U.S. Drug Policy." *Oregon Law Review* 91 (4): 1069–95.

Boettke, Peter J., Christopher J. Coyne, and Peter T. Leeson. 2008. "Institutional Stickiness and the New Development Economics." *American Journal of Economics and Sociology* 67 (2): 331–58.

Brennan, Geoffrey, and Loren E. Lomasky. 1993. *Democracy and Decision: The Pure Theory of Electoral Preference*. Cambridge, UK: Cambridge University Press.

Brown, Gardner M., and Jason F. Shogren. 1998. "Economics of the Endangered Species Act." *The Journal of Economic Perspectives* 12 (3): 3–20.

Buchanan, James M. 1999a. "Politics without Romance: A Sketch of Positive Public Choice Theory and Its Normative Implications." In *The Logical Foundations of Constitutional Liberty*, 1:45–59. The Collected Works of James M. Buchanan. Indianapolis, IN: Liberty Fund.

Buchanan, James M. 1999b. "What Should Economists Do?" In *The Logical Foundations of Constitutional Liberty*, 1:28–42. The Collected Works of James M. Buchanan. Indianapolis, IN: Liberty Fund.

Caplan, Bryan. 2007. *The Myth of the Rational Voter: Why Democracies Choose Bad Policies*. Princeton, NJ: Princeton University Press.

Carlson, Ann E. 2012. "Designing Effective Climate Policy: Cap-and-Trade and Complementary Policies." *Harvard Journal on Legislation* 49 (2): 207–48.

Carson, Jennifer Varriale. 2014. "Counterterrorism and Radical Eco-Groups: A Context for Exploring the Series Hazard Model." *Journal of Quantitative Criminology* 30 (3): 485–504.

Carson, Jennifer Varriale, Gary LaFree, and Laura Dugan. 2012. "Terrorist and Non-Terrorist Criminal Attacks by Radical Environmental and Animal Rights Groups in the United States, 1970–2007." *Terrorism and Political Violence* 24 (2): 295–319.

Ceballos, Gerardo, Paul R. Ehrlich, Anthony D. Barnosky, Andrés García, Robert M. Pringle, and Todd M. Palmer. 2015. "Accelerated Modern Human–Induced Species Losses: Entering the Sixth Mass Extinction." *Science Advances* 1 (5): e1400253.

Chapman, Laura A., and Piran C. L. White. 2020. "Stakeholder Perspectives on the Value and Challenges of Private Rhinoceros Ownership in South Africa." *Human Dimensions of Wildlife* 25 (2): 187–97.

Chermak, Steven M., Joshua Freilich, Celinet Duran, and William S. Parkin. 2013. "An Overview of Bombing and Arson Attacks by Environmental and Animal Rights Extremists in the United States, 1995-2010." Final Report to the Resilient Systems Division, Science and Technology Directorate, US Department of Homeland Security. https://www.dhs.gov/sites/default/files/publications/OPSR_TP_TEVUS_Bombing-Arson-Attacks_Environmental-Animal%20Rights-Extremists_1309-508.pdf.

Chu, Cindy. 2009. "Thirty Years Later: The Global Growth of ITQs and Their Influence on Stock Status in Marine Fisheries." *Fish and Fisheries (Oxford, England)* 10 (2): 217–30.

Chu, Jingjie, James L. Anderson, Frank Asche, and Lacey Tudur. 2010. "Stakeholders' Perceptions of Aquaculture and Implications for Its Future: A Comparison of the U.S.A. and Norway." *Marine Resource Economics* 25 (1): 61–76.

Clements, Hayley S., Dave Balfour, and Enrico Di Minin. 2023. "Importance of Private and Communal Lands to Sustainable Conservation of Africa's Rhinoceroses." *Frontiers in Ecology and the Environment* 21 (3): 140–47.

Coase, Ronald H. 1960. "The Problem of Social Cost." *The Journal of Law & Economics* 3: 1–44.

Cohen, Maurie J. 2011. "Is the UK Preparing for 'War'? Military Metaphors, Personal Carbon Allowances, and Consumption Rationing in Historical Perspective." *Climatic Change* 104 (2): 199–222.

Cole, Daniel H. 2015. "Advantages of a Polycentric Approach to Climate Change Policy." *Nature Climate Change* 5 (2): 114–18.

Cowie, Robert H., Philippe Bouchet, and Benoît Fontaine. 2022. "The Sixth Mass Extinction: Fact, Fiction or Speculation?" *Biological Reviews of the Cambridge Philosophical Society* 97 (2): 640–63.

Dees, J. Gregory, Peter Economy, and Jed Emerson. 2001. *Enterprising Nonprofits: A Toolkit for Social Entrepreneurs*. Wiley Nonprofit Law, Finance, and Management Series. New York: Wiley.

Delina, Laurence L. 2016. *Strategies for Rapid Climate Mitigation: Wartime Mobilisation as a Model for Action?* Routledge Advances in Climate Change Research. Abingdon, Oxon: Routledge, Earthscan.

Desrochers, Pierre, and Joanna Szurmak. 2020a. "Population Growth and the Governance of Complex Institutions: People Are More than Mouths to Feed." In *The Environmental Optimism of Elinor Ostrom*, 91–147. Logan: Center for Growth and Opportunity at Utah State University.

Desrochers, Pierre, and Joanna Szurmak. 2020b. "The Environmental Benefits of Long-Distance Trade: Insights from the History of By-Product Development." In *The*

Environmental Optimism of Elinor Ostrom, 173–208. Logan: Center for Growth and Opportunity at Utah State University.

Dorman, Maxwell T., Aaron L. Strong, and Nicola Ulibarri. 2022. "Coalitions in Climate Mitigation Policy Re-Design Processes: The Case of the Regional Greenhouse Gas Initiative." *Environmental Science & Policy* 127 (January): 38–47.

Druedahl, Louise C., Timo Minssen, and W. Nicholson Price. 2021. "Collaboration in Times of Crisis: A Study on COVID-19 Vaccine R&D Partnerships." *Vaccine* 39 (42): 6291–95.

Duncan, Emily, Alesandros Glaros, Dennis Z. Ross, and Eric Nost. 2021. "New but for Whom? Discourses of Innovation in Precision Agriculture." *Agriculture and Human Values* 38 (4): 1181–99.

Ellickson, Robert C. 1991. *Order without Law: How Neighbors Settle Disputes*. Cambridge, MA: Harvard University Press.

Erdogan, Seyfettin, Ayfer Gedikli, Emrah İsmail Çevik, and Mehmet Akif Öncü. 2022. "Does Military Expenditure Impact Environmental Sustainability in Developed Mediterranean Countries?" *Environmental Science and Pollution Research International* 29 (21): 31612–30.

Eregha, Perekunah B., Xuan Vinh Vo, and Solomon Prince Nathaniel. 2022. "Military Spending, Financial Development, and Ecological Footprint in a Developing Country: Insights from Bootstrap Causality and Maki Cointegration." *Environmental Science and Pollution Research International* 29 (55): 83945–55.

Eze, Chidiebere Emmanuel, Rex Asibuodu Ugulu, Samuel Ikechukwu Egwunatum, and Imoleayo Abraham Awodele. 2021. "Green Building Materials Products and Service Market in the Construction Industry." *Journal of Engineering, Project, and Production Management* 11 (2): 89–101.

Feng, Sida, and Christopher L. Magee. 2020. "Technological Development of Key Domains in Electric Vehicles: Improvement Rates, Technology Trajectories and Key Assignees." *Applied Energy* 260 (February): 114264.

Ferreira, Sam M., Michele Pfab, and Mike Knight. 2014. "Management Strategies to Curb Rhino Poaching: Alternative Options Using a Cost-Benefit Approach." *South African Journal of Science* 110 (5–6): 1–8.

Flusberg, Stephen J., Teenie Matlock, and Paul H. Thibodeau. 2017. "Metaphors for the War (or Race) against Climate Change." *Environmental Communication* 11 (6): 769–83.

Geertz, Clifford. 1973. "Thick Description: Toward an Interpretive Theory of Culture." In *The Interpretation of Cultures*, 3–30. New York, NY: Basic Books.

Geloso, Vincent. 2022. "Statogenic Climate Change? Julian Simon and Institutions." *The Review of Austrian Economics* 35 (3): 343–58.

Gibbs, Katherine E., and David J. Currie. 2012. "Protecting Endangered Species: Do the Main Legislative Tools Work?" *PloS One* 7 (5): e35730.

Gillanders, Robert, and Lisa Werff. 2022. "Corruption Experiences and Attitudes to Political, Interpersonal, and Domestic Violence." *Governance (Oxford)* 35 (1): 167–85.

Grube, Laura E., and Virgil Henry Storr, eds. 2015. *Culture and Economic Action*. New Thinking in Political Economy. Cheltenham, UK: Edward Elgar Publishing.

Haeffele, Stefanie, and Virgil Henry Storr. 2020. *Bottom-Up Responses to Crisis*. Mercatus Studies in Political and Social Economy. Cham, Switzerland: Palgrave Macmillan.

Hayek, Friedrich A. 1945. "The Use of Knowledge in Society." *The American Economic Review* 35 (4): 519–30.

Hayek, Friedrich A. 1976. *Law, Legislation and Liberty, Volume 2: The Mirage of Social Justice*. Chicago, IL: University of Chicago Press.

Hayek, Friedrich A. 2014. "The Results of Human Action but Not of Human Design (1967)." In *The Collected Works of F. A. Hayek, Volume 15: The Market and Other Orders*, 293–303. Chicago, IL: University of Chicago Press.

Hernandez-Cortes, Danae, and Kyle C. Meng. 2023. "Do Environmental Markets Cause Environmental Injustice? Evidence from California's Carbon Market." *Journal of Public Economics* 217 (January): 104786.

Higgs, Robert. 2012. *Crisis and Leviathan: Critical Episodes in the Growth of American Government*. 25th anniversary edition. Oakland, CA: Independent Institute.

Hodson, Dermot, Uwe Puetter, Sabine Saurugger, and John Peterson. 2022. *The Institutions of the European Union*. 5th ed. Oxford, UK: Oxford University Press.

Holcombe, Randall G. 2018. *Political Capitalism: How Economic and Political Power Is Made and Maintained*. Cambridge, UK: Cambridge University Press.

Holley, Cameron, and Darren Sinclair. 2016. "Rethinking Australian Water Law and Governance: Successes, Challenges and Future Directions." *Environmental and Planning Law Journal* 33 (4): 275–83.

Hooks, Gregory, and Chad L. Smith. 2005. "Treadmills of Production and Destruction: Threats to the Environment Posed by Militarism." *Organization & Environment* 18 (1): 19–37.

Hoshino, Eriko, Ingrid van Putten, Sean Pascoe, and Simon Vieira. 2020. "Individual Transferable Quotas in Achieving Multiple Objectives of Fisheries Management." *Marine Policy* 113 (March): 103744.

Huang, Yimeng, and Ju Li. 2022. "Key Challenges for Grid-Scale Lithium-Ion Battery Energy Storage." *Advanced Energy Materials* 12 (48): 2202197.

Hundley, Norris, Jr. 2009. *Water and the West: The Colorado River Compact and the Politics of Water in the American West*. 2nd ed. Berkeley: University of California Press.

Innes, Robert. 2000. "The Economics of Takings and Compensation When Land and Its Public Use Value Are in Private Hands." *Land Economics* 76 (2): 195–212.

IPCC. 2022. *Climate Change 2022: Impacts, Adaptation and Vulnerability*. Cambridge, UK: Cambridge University Press. https://www.ipcc.ch/report/ar6/wg2/downloads/report/IPCC_AR6_WGII_FullReport.pdf.

Jacobsen, Peter, and Louis Rouanet. 2022. "Economists versus Engineers: Two Approaches to Environmental Problems." *The Review of Austrian Economics* 35 (3): 359–81.

Johnson, Timothy, and Ameya Joshi. 2018. "Review of Vehicle Engine Efficiency and Emissions." *SAE International Journal of Engines* 11 (6): 1307–30.

Kebreab, Ermias, dir. 2022. *What Seaweed and Cow Burps Have to Do with Climate Change | Ermias Kebreab | TED Countdown*. TED. https://www.youtube.com/watch?v=fCpKG1QJC14.

Kester, Johannes, and Benjamin K. Sovacool. 2017. "Torn between War and Peace: Critiquing the Use of War to Mobilize Peaceful Climate Action." *Energy Policy* 104: 50–55.

Keynes, John Maynard, Elizabeth Johnson, and Donald Moggridge, eds. 1978. "Editor and Referee." In *The Collected Writings of John Maynard Keynes: Volume 12: Economic Articles and Correspondence: Investment and Editorial*, 12: 784–868. The Collected Writings of John Maynard Keynes. London: Royal Economic Society.

Kirzner, Israel M. 1973. *Competition and Entrepreneurship*. Chicago, IL: University of Chicago Press.

Kirzner, Israel M. 1997. "Entrepreneurial Discovery and the Competitive Market Process: An Austrian Approach." *Journal of Economic Literature* 35 (1): 60–85.

Knight, Frank Hyneman. 1985. *Risk, Uncertainty and Profit*. Midway Reprint ed. Chicago, IL: University of Chicago Press.

Lavoie, Don. 1986. "The Market as a Procedure for Discovery and Conveyance of Inarticulate Knowledge." *Comparative Economic Studies* 28 (1): 1–19.

Lavoie, Don. 2015. *Rivalry and Central Planning*. Arlington, VA: Mercatus Center at George Mason University.

Lavoie, Don. 2016. *National Economic Planning: What Is Left?* Arlington, VA: Mercatus Center at George Mason University.

Leeson, Peter T. 2009. *The Invisible Hook: The Hidden Economics of Pirates*. Princeton, NJ: Princeton University Press.

Leeson, Peter T. 2017. *WTF?!: An Economic Tour of the Weird*. Stanford, CA: Stanford University Press.

Lessmann, Christian, and Niklas Kramer. 2024. "The Effect of Cap-and-Trade on Sectoral Emissions: Evidence from California." *Energy Policy* 188 (May): 114066.

Levy, David M., and Sandra J. Peart. 2001. "The Secret History of the Dismal Science. Part I. Economics, Religion and Race in the 19th Century." The Library of Economics and Liberty. https://www.econlib.org/library/Columns/LevyPeartdismal.html.

Lofthouse, Jordan K. 2020. "Self-Governance, Polycentricity, and Environmental Policy." In *The Environmental Optimism of Elinor Ostrom*, 31–60. Logan: Center for Growth and Opportunity at Utah State University.

Lofthouse, Jordan K., and Roberta Q. Herzberg. 2023. "The Continuing Case for a Polycentric Approach for Coping with Climate Change." *Sustainability (Basel, Switzerland)* 15 (4): 3770.

Lofthouse, Jordan K., and Megan E. Jenkins. 2020. "Cooperation or Conflict: Two Approaches to Conservation." In *Regulation and Economic Opportunity: Blueprints for Reform*, 247–76. Logan: Center for Growth and Opportunity at Utah State University.

Lueck, Dean, and Jeffrey A. Michael. 2003. "Preemptive Habitat Destruction under the Endangered Species Act." *The Journal of Law & Economics* 46 (1): 27–60.

Lukanov, Boris R., and Elena M. Krieger. 2019. "Distributed Solar and Environmental Justice: Exploring the Demographic and Socio-Economic Trends of Residential PV Adoption in California." *Energy Policy* 134 (November): 110935.

Malm, Andreas. 2020. *Corona, Climate, Chronic Emergency: War Communism in the Twenty-First Century*. Coronavirus Pamphlet Series. London: Verso.

Mann, Michael E. 2021. *The New Climate War: The Fight to Take Back Our Planet*. 1st ed. New York: PublicAffairs.

Matos, Catarina R., Patrícia P. Silva, and Júlio F. Carneiro. 2022. "Overview of Compressed Air Energy Storage Projects and Regulatory Framework for Energy Storage." *Journal of Energy Storage* 55 (November): 105862.

McCloskey, Deirdre. 1998. "The So-Called Coase Theorem." *Eastern Economic Journal* 24 (3): 367–71.

McCloskey, Deirdre N. 2010. *Bourgeois Dignity: Why Economics Can't Explain the Modern World*. Chicago, IL: University of Chicago Press.

McCloskey, Deirdre Nansen. 2019. *Why Liberalism Works: How True Liberal Values Produce a Freer, More Equal, Prosperous World for All*. New Haven, CT: Yale University Press.

Messmer, Terry A., Lorien Belton, David Dahlgren, S. Nicole Frey, Michel Kohl, and Ray Ann Hart. 2019. "Utah's Adaptive Resources Management Greater Sage-Grouse Local Working Groups 2018 Annual Report." Jack H. Berryman Institute, Department of Wildland Resources, and Utah State University Extension, Logan. https://extension.usu.edu/utahcbcp/files/2018LWGAnnualReportMarch2019.pdf.

Miron, Jeffrey A. 2004. *Drug War Crimes: The Consequences of Prohibition*. Oakland, CA: The Independent Institute.

Mitchell, Matthew D. 2014. *The Pathology of Privilege: The Economic Consequences of Government Favoritism*. Arlington, VA: Mercatus Center at George Mason University.

Mokyr, Joel. 2016. *A Culture of Growth: The Origins of the Modern Economy*. The Graz Schumpeter Lectures. Princeton, NJ: Princeton University Press.

Murray, Brian C., and Peter T. Maniloff. 2015. "Why Have Greenhouse Emissions in RGGI States Declined? An Econometric Attribution to Economic, Energy Market, and Policy Factors." *Energy Economics* 51 (September): 581–89.

Niskanen, William A. 1971. *Bureaucracy and Representative Government*. New York City, NY: Routledge.

North, Douglass C. 1990. *Institutions, Institutional Change, and Economic Performance*. New York City, NY: Cambridge University Press.

Olson, James M. 2006. "Navigating the Great Lakes Compact: Water, Public Trust, and International Trade Agreements." *Michigan State Law Review* 2006: 1103–40.

Ostrom, Elinor. 1990. *Governing the Commons: The Evolution of Institutions for Collective Action*. New York City, NY: Cambridge University Press.

Ostrom, Elinor. 1996. "Crossing the Great Divide: Coproduction, Synergy, and Development." *World Development* 24 (6): 1073–87.

Ostrom, Elinor. 2005. *Understanding Institutional Diversity*. Princeton, NJ: Princeton University Press.

Ostrom, Elinor. 2010. "Beyond Markets and States: Polycentric Governance of Complex Economic Systems." *American Economic Review* 100 (3): 641–72.

Ostrom, Elinor. 2011. "A Multi-Scale Approach to Coping with Climate Change and Other Collective Action Problems." *Leviathan (Düsseldorf)* 39 (3): 447–58.

Ostrom, Elinor. 2012. "Nested Externalities and Polycentric Institutions: Must We Wait for Global Solutions to Climate Change before Taking Actions at Other Scales?" *Economic Theory* 49 (2): 353–69.

Ostrom, Elinor. 2014. "A Polycentric Approach for Coping with Climate Change." *Annals of Economics and Finance* 15 (1): 97–134.

Ostrom, Vincent. 1973. *The Intellectual Crisis in American Public Administration*. Tuscaloosa, AL: University of Alabama Press.

Ostrom, Vincent. 1997. *The Meaning of Democracy and the Vulnerability of Democracies: A Response to Tocqueville's Challenge*. Ann Arbor, MI: University of Michigan Press.

Ostrom, Vincent, Charles M. Tiebout, and Robert Warren. 1961. "The Organization of Government in Metropolitan Areas: A Theoretical Inquiry." *The American Political Science Review* 55 (4): 831–42.

Pooley, Gale, and Marian Tupy. 2020. "Luck or Insight? The Simon–Ehrlich Bet Re-examined." *Economic Affairs* 40 (2): 277–80.

Prieto, Manuel. 2022. "Indigenous Resurgence, Identity Politics, and the Anticommodification of Nature: The Chilean Water Market and the Atacameño People." *Annals of the American Association of Geographers* 112 (2): 487–504.

Qayyum, Unbreen, Sohail Anjum, and Samina Sabir. 2021. "Armed Conflict, Militarization and Ecological Footprint: Empirical Evidence from South Asia." *Journal of Cleaner Production* 281: 125299.

Rajaeifar, Mohammad Ali, Oliver Belcher, Stuart Parkinson, Benjamin Neimark, Doug Weir, Kirsti Ashworth, Reuben Larbi, and Oliver Heidrich. 2022. "Decarbonize the Military—Mandate Emissions Reporting." *Nature (London)* 611 (7934): 29–32.

Read, Leonard E. 1958. *I, Pencil: My Family Tree*. Irvington-on-Hudson, NY: Foundation for Economic Education.

Redford, Audrey, and Angela K. Dills. 2021. "The Political Economy of Drug and Alcohol Regulation during the COVID-19 Pandemic." *Southern Economic Journal* 87 (4): 1175–1209.

Reyes, Diana C., Jennifer Meredith, Leah Puro, Katherine Berry, Richard Kersbergen, Kathy J. Soder, Charlotte Quigley, et al. 2023. "Maine Organic Dairy Producers' Receptiveness to Seaweed Supplementation and Effect of Chondrus Crispus on Enteric Methane Emissions in Lactating Cows." *Frontiers in Veterinary Science* 10: 1153097.

Robbins, Lionel Robbins. 1935. *An Essay on the Nature and Significance of Economic Science*. 2nd ed., rev, and extended. London: Macmillan.

Rowland, Mary M., Michael J. Wisdom, Lowell H. Suring, and Cara W. Meinke. 2006. "Greater Sage-Grouse as an Umbrella Species for Sagebrush-Associated Vertebrates." *Biological Conservation* 129 (3): 323–35.

Rubino, Elena C., and Elizabeth F. Pienaar. 2020. "Rhinoceros Ownership and Attitudes towards Legalization of Global Horn Trade within South Africa's Private Wildlife Sector." *Oryx* 54 (2): 244–51.

Schumpeter, Joseph A. 2008. *Capitalism, Socialism and Democracy*. 3rd ed. New York City, NY: Harper Perennial Modern Thought.

Scott, J. Michael, Dale D. Goble, John A. Wiens, David S. Wilcove, Michael Bean, and Timothy Male. 2005. "Recovery of Imperiled Species under the Endangered Species Act: The Need for a New Approach." *Frontiers in Ecology and the Environment* 3 (7): 383–89.

Shah, Parth J., and Vidisha Maitra, eds. 2005. *Terracotta Reader: A Market Approach to the Environment*. New Delhi, India: Academic Foundation in association with Centre for Civil Society, New Delhi.

Sharp, Gene. 1973. *The Politics of Nonviolent Action*. Extending Horizons Books. Boston: P. Sargent Publisher.

Shockley, Gordon E., Roger R. Stough, Kingsley E. Haynes, and Peter M. Frank. 2006. "Toward a Theory of Public Sector Entrepreneurship." *International Journal of Entrepreneurship and Innovation Management* 6 (3): 205–23.

Simon, Julian Lincoln. 1981. *The Ultimate Resource*. 1st ed. Princeton, NJ: Princeton University Press.

Smith, Adam. 1982. *An Inquiry into the Nature and Causes of the Wealth of Nations*. Vol. 1. Indianapolis, IN: Liberty Fund.

Smith, Adam C., and Bruce Yandle. 2014. *Bootleggers and Baptists: How Economics Forces and Moral Persuasion Interact to Shape Regulatory Politics*. Washington, DC: Cate Institute.

Smith, Chad L., Gregory Hooks, and Michael Lengefeld. 2015. "The War on Drugs in Colombia: The Environment, the Treadmill of Destruction and Risk-Transfer Militarism." *Journal of World-Systems Research* 20 (2): 185–206.

Smith, Chad L., and Michael R. Lengefeld. 2020. "The Environmental Consequences of Asymmetric War: A Panel Study of Militarism and Carbon Emissions, 2000–2010." *Armed Forces and Society* 46 (2): 214–37.

Stopps, Helen, and Marianne F. Touchie. 2021. "Residential Smart Thermostat Use: An Exploration of Thermostat Programming, Environmental Attitudes, and the Influence of Smart Controls on Energy Savings." *Energy and Buildings* 238 (May): 110834.

Storr, Virgil Henry, Stefanie Haeffele-Balch, and Laura E. Grube. 2015. *Community Revival in the Wake of Disaster: Lessons in Local Entrepreneurship*. New York City, NY: Palgrave Macmillan.

Swift, Byron. 2001. "Command Without Control: Why Cap-and-Trade Should Replace Rate Standards for Regional Pollutants." *Environmental Law Reporter* 31 (3): 10330.

't Sas-Rolfes, Michael. n.d. "Saving African Rhinos: A Market Success Story." PERC. https://www.perc.org/wp-content/uploads/2011/08/Saving-African-Rhinos-final.pdf.

Tarko, Vlad. 2017. *Elinor Ostrom: An Intellectual Biography*. New York City, NY: Rowman and Littlefield.

Taylor, Martin F. J., Kieran F. Suckling, and Jeffrey J. Rachlinski. 2005. "The Effectiveness of the Endangered Species Act: A Quantitative Analysis." *Bioscience* 55 (4): 360–67.

Tuladhar, Sugandha D., Sebastian Mankowski, and Paul Bernstein. 2014. "Interaction Effects of Market-Based and Command-and-Control Policies." *The Energy Journal (Cambridge, Mass.)* 35 (1_suppl): 61–88.

Tullock, Gordon. 2004a. "The Transitional Gains Trap." In *Virginia Political Economy*. Vol. 1. The Selected Works of Gordon Tullock. Indianapolis, IN: Liberty Fund.

Tullock, Gordon. 2004b. "The Welfare Costs of Tariffs, Monopolies, and Theft." In *Virginia Political Economy*, 169–79. The Selected Works of Gordon Tullock. Indianapolis, IN: Liberty Fund.

Tullock, Gordon. 2005. *Bureaucracy*. Vol. 6. The Selected Works of Gordon Tullock. Indianapolis, IN: Liberty Fund.

US Fish and Wildlife Service. 2017. "Critical Habitat: What Is It?" https://www.fws.gov/sites/default/files/documents/critical-habitat-fact-sheet.pdf.

US Fish and Wildlife Service. 2024a. "Arctic National Wildlife Refuge." https://www.fws.gov/refuge/arctic/about-us.

US Fish and Wildlife Service. 2024b. "Conservation Benefit Agreements." https://www.fws.gov/service/conservation-benefit-agreements.

US Fish and Wildlife Service. 2024c. "Habitat Conservation Plans." https://www.fws.gov/service/habitat-conservation-plans.

US Fish and Wildlife Service. 2024d. "Mille Lacs National Wildlife Refuge." https://www.fws.gov/refuge/mille-lacs#:~:text=Mille%20Lacs%20National%20Wildlife%20Refuge%20was%20established%20in%201915%20and,Refuge%20System%20at%200.57%20acres.

US Fish and Wildlife Service. 2024e. "Recovery Land Acquisition Grants." https://www.fws.gov/service/recovery-land-acquisition-grants.

US Fish and Wildlife Service. 2024f. "Safe Harbor Agreements." https://www.fws.gov/service/safe-harbor-agreements.

Wagner, Richard E. 2016. *Politics as a Peculiar Business*. Cheltenham, UK: Edward Elgar Publishing.

Wang, Stephen W., and Maxwell K. Hsu. 2016. "Airline Co-Branded Credit Cards—An Application of the Theory of Planned Behavior." *Journal of Air Transport Management* 55 (August): 245–54.

Wheeler, Sarah Ann, Adam Loch, Lin Crase, Mike Young, and R. Quentin Grafton. 2017. "Developing a Water Market Readiness Assessment Framework." *Journal of Hydrology* 552 (September): 807–20.

Yan, Jingchi. 2021. "The Impact of Climate Policy on Fossil Fuel Consumption: Evidence from the Regional Greenhouse Gas Initiative (RGGI)." *Energy Economics* 100 (August): 105333.

Young, Michael D. 2015. "Unbundling Water Rights as a Means to Improve Water Markets in Australia's Southern Connected Murray-Darling Basin." In *Use of Economic Instruments in Water Policy: Insights from International Experience*, Global Issues in Water Policy, vol. 14: 279–99. Switzerland: Springer International Publishing AG.

Young, Michael D., and Jim C McColl. 2009. "Double Trouble: The Importance of Accounting for and Defining Water Entitlements Consistent with Hydrological Realities." *The Australian Journal of Agricultural and Resource Economics* 53 (1): 19–35.

Zhao, Maruge, Tiho Ancev, and R. Willem Vervoort. 2024. "Water Market Functionality: Evidence from the Australian Experience." *Water Resources Research* 60 (2).

INDEX

Africa 8–9, 131, 139–48
agriculture
    and climate change 11, 175–6, 179, 180–3
    and conservation 136, 142, 163, 165–6
    and entrepreneurship 57
    and Neo-Malthusianism 120–2
    and public choice economics 84
    and tradeoffs 27–9
airplanes 1, 21, 54, 73
American Prairie Reserve 10, 153–9, 194
Anderson, Terry L. 5, 8, 102, 118, 161
Australia 10–11, 20, 91, 161–9, 181–2

Bastiat, Frédéric 69–70
bison 44–5, 151, 153–9
black markets 17
Bootleggers and Baptists 82–3, 104–6
broken window fallacy 69–70
Buchanan, James M. 3, 4, 77
bureaucracy
    and conservation 131–8, 173–4
    public administration 40, 44, 89, 161
    and public choice economics 7, 75, 77, 80–1, 84–7, 104

California 27, 94, 117, 169, 183–6
cap-and-trade systems 47, 110–14, 183–7
Carlson, Ann E. 113–14
catallaxy 3, 153–9
central economic planning 38, 58–9, 176–8
civil society
    and American Prairie Reserve 153–9
    and conservation 114–17, 128–9, 131, 147–8
    and governance 90–92, 95–7, 102–3, 178, 189
    informal institutions 38
    peaceful cooperation 192–3
    private sphere 76
    and entrepreneurship 8, 107
climate change
    activism 1–2
    and entrepreneurship 58, 125
    and opportunity costs 27
    and polycentricity 11–12, 138, 175–90
    and public policies 103
    and water scarcity 161, 163, 170
Coase Theorem 49–50, 114–16, 153–9
Coase, Ronald 4, 49–50, 116
coercion
    and free market environmentalism 107, 119
    foreign intervention 39
    government power 8, 45, 75–6, 80–1, 122
    and liberalism 194
common-pool resources 46–8, 102, 121
competition
    market competition 59, 63–4
    and polycentricity 11–12, 91–3, 96, 177–8
    and public choice economics 80–1, 83, 86, 103
    and unintended consequences 70
conflict
    and climate change 177
    and cooperation 38, 49, 90
    and peace 12, 191–4
    and politics 10, 102–3, 109, 111, 120, 132, 159
    and water scarcity 11, 161, 168, 174
Congress 22, 40, 44, 83–5, 132, 149–51, 175

conservation
    and free market environmentalism
        103, 106, 114, 119
    land conservation 149–59, 194
    and property rights 45–6
    species conservation 8–10, 131–48
    water conservation 94, 163, 169–74
constraints
    definition of 20–1
    and public choice economics 76–7,
        83–7, 104
    legal constraints 90, 153
    and free market environmentalism 108
cooperation
    and American Prairie Reserve 153–9
    and institutions 37–8, 115–16, 138
    and peace 168, 192–3
    and polycentricity 90–4, 177–8
    and positive-sum games 60, 103, 109
    and public choice economics 80
    and tragedy of the commons 46
coproduction 92, 94–5, 131–8, 177
creative destruction 56
culture 4, 17, 38–9, 45, 107, 120, 126–7

Defenders of Wildlife 114–16
democracy 12, 75, 77, 89–91, 195–6
disasters 66, 119–20
division of labor 51–3

Ehrlich, Paul 121–5
Ellickson, Robert 115, 117
endangered species
    and American Prairie Reserve 153–9
    Endangered Species Act 9, 40, 131–8,
        172
    greater sage-grouse 9, 131–8
    normative and positive analysis 16
    and property rights 45–6
    rhinos 8–9, 139–48
    and tradeoffs 27
    wolves 114–16
entrepreneurship
    and the market process 45, 56–9, 86
    and supply and demand 63–4

problem-solving 7–8, 124–8
    and water scarcity 161, 166, 173, 174
    and free market environmentalism 10,
        12, 101–2, 110, 112, 114–20,
        142–3, 146, 194–5
environmental activism 1–2, 153–9, 161,
    176, 192
Environmental Kuznets Curve 126–8
equilibrium 63–8
experimentation
    and conservation 153–9
    policy experimentation 128–9, 136,
        172–3
    and polycentricity 12, 95–7, 177–8
    scientific experimentation 181
externalities
    and American Prairie Reserve 10,
        155–9
    definition of 48–9
    and politics 27, 81
    pollution 112
    and property rights 7, 102–3

fisheries 46–7
free market environmentalism 5, 8, 102–4,
    107–10, 117–20, 153–9, 169–74

greater sage-grouse 9, 131–8, 147
greenhouse gas emissions 2, 11, 16, 103,
    105, 113, 175–6

Hardin, Garrett 46, 121
Hayek, Friedrich A. 3, 4, 18, 55

incentives
    and American Prairie Reserve 10, 151,
        156–9
    definition of 20
    and free market environmentalism 8,
        108–15
    and innovation 125–6
    and institutions 38, 128–9
    and market process 53, 56, 59, 64,
    and property rights 7, 43, 45–6
    and peace 192

and polycentricity 92–7, 176–7, 188–9
and public choice economics 76–87, 104,
and species conservation 9, 131–48,
and water scarcity 161, 163, 168–9, 170, 174
innovation
    and market process 45, 50, 56–7, 59, 66
    and polycentricity 92–3, 116, 178–90
    and entrepreneurship 7, 101, 103, 107–10, 118–20, 123–8
    and unintended consequences 17
    and opportunity costs 70
    and liberty 195
    and conservation 10, 134–6, 146, 153–9
institutions
    and environmental problems 6–8, 101–2, 107–10, 114–17, 120, 125
    and public choice economics 77, 87
    and feedback 58
    and property rights 43, 45
    definition of 37–41
    and Environmental Kuznets Curve 127–9
    and conservation 131, 139, 141–5
    and peace 192–4
    and water scarcity 163–5, 169–74

judges 86, 104

Kirzner, Israel 5, 56–9
knowledge
    dispersed 116, 118, 137–8, 195
    and institutions 38, 86, 174
    and the market process 52–55, 59, 108, 112–13, 120, 123–4
    and politics 104, 113, 176–7
    and polycentricity 12, 95–7, 188–90
Kyoto Protocol 105–6

Lavoie, Don 59
Leal, Donald R. 5, 8, 102, 118, 161
life cycle assessment 28

Malthus, Thomas Robert 121–2
Marginal Revolution 31
marginalism 6, 31–5, 76, 78, 112
market process
    and entrepreneurship 118, 124, 128
    explanation of 51–60, 86
    and peace 192
    and water 161–9
Marx, Karl 31
methodological individualism 21–2
Montana 10, 44, 115, 153–9, 171, 173, 194

national parks 10, 44–5, 114–16, 141, 149–53
negative-sum games 8, 38, 60, 80, 101–4, 106, 128–9, 174
nirvana fallacy 8
Niskanen, William A. 5, 85–6
normative analysis 16–17, 26, 89–90, 142, 174, 185, 191–7
North, Douglass 4, 37–8

opportunity cost
    definition of 26–7
    and prices 55, 162, 167, 174
    and public choice economics 80, 82, 104, 176
    and conservation 142, 146–7
    and the broken window fallacy 70
Ostrom, Elinor 4, 47–8, 89–91, 102, 116–17, 121, 176–7, 190
Ostrom, Vincent 89–91, 195–6
overpopulation 119–25

philanthropy 10, 35, 115, 147, 153–4
philosophy 4, 15–17, 26, 75–6, 185, 192–4
pirates 19, 37
plastics 23, 28, 34
policymaking
    cap-and-trade systems 113–14, 185,
    conservation policy 9–10, 131–8, 140, 146–7, 150–2, 155, 161, 166, 171–4
    and free market environmentalism 102–4, 109–11

political entrepreneurship 57–8,
    and polycentricity 11–12, 89–90, 93,
        95–7, 116, 176, 187–8
    process of policymaking 40
    and public choice economics 77–8,
        82–3
    and self-interest 22
    and unintended consequences 17, 28,
        70–1
politicians 7, 22–3, 77–84, 86–7, 89, 106
pollution
    and cap-and-trade systems 110–14,
        184–5
    and Coase Theorem 50,
    and Environmental Kuznets Curve 127
    and marginalism 33,
    and Neo-Malthusianism 122
    and property rights 7, 47–8, 102–3
polycentricity
    characteristics of 7, 89–97, 116, 196
    and climate change 11–12, 176–90
    and conservation 131–8, 171–3
positive analysis 16–17, 89–90
positive-sum game 7–8, 38, 59–60, 101–2,
    106–7, 109, 114–15, 128–9, 153–9,
    191–2, 197
prices
    function of 54–59, 61–8, 86, 92, 110,
        183
    price controls 17, 65–68, 70–1
    Simon–Ehrlich Bet 123, 125,
    and species conservation 140, 145–6
    and water markets 11, 161–9, 174
profit and loss 56–9, 71, 86–7, 105, 112,
    119, 140, 143, 156, 158,
property rights
    explanation of 7–10, 43–50
    and free market environmentalism
        101–10, 115, 118–19, 120, 129
    and land conservation 153–9
    in the market process 53–4
    and public choice economics 75
    and species conservation 138–48
    and water markets 161–74

public choice economics 6–7, 75–87, 104,
    120, 128, 158–9, 176, 188

rationality
    and endangered species 133, 138, 142,
        144
    and entrepreneurship 56,
    explanation of 17–23
    and public choice economics 76,
        78–80, 83–6
    and tradeoffs 26,
    and tragedy of the commons 46
Read, Leonard E. 72–3
recycling 34–5, 57
rent-seeking 80–3, 103–6, 109, 120, 174,
    177
resilience 12, 96, 178
rhinos 8–9, 139–48, 194
risk 54, 71, 118, 143–6,

scarcity 3, 5, 10–11, 25–6, 55, 122, 125,
    161–74
Schumpeter, Joseph 5, 56
self-interest
    definition 5
    and cap-and-trade systems 112
    and conservation 133, 139–48
    and environmental policy 22–3
    and free market environmentalism
        108,
    and market process 56, 65
    and public choice economics 7,
        76–87
    and spontaneous order 72–4
    and the tragedy of the commons 46
    and peace 192
Simon, Julian 5, 102, 121–8
Smith, Adam 3–5, 15, 31, 51–2, 72
specialization 4, 51–3, 118,
spontaneous order 71–4, 97, 166–7, 177,
    192–3
subjectivity 18, 25–6, 31–2, 34, 53, 60
supply and demand 6, 17, 61–68, 110, 122,
    126, 145–6, 161–9

Tocqueville, Alexis de 90
tradeoffs
    and cap-and-trade systems 111–13, 184–8
    and conservation 142–3, 147, 150, 152, 163
    definition of 5–6, 25–9
    of fossil fuels 175
    and marginalism 33
    and private decisions 76
    and property rights 44, 47
    and public policies 104, 107–8
    and voters 78
tragedy of the commons 46–8, 102, 121
transaction costs 49–50, 109–10, 113, 162, 165–6
Tullock, Gordon 4, 85–6
Turner, Ted 44–5

U.S. Fish and Wildlife Service 9, 40, 131–8, 149, 152–5, 172,
uncertainty 11, 29, 54, 106, 138

unintended consequences
    and the broken window fallacy 69–70
    and climate change 176, 188–9
    and endangered species 132–3, 147–8,
    negative 17, 70–1, 103–4, 107–8, 111, 113, 116, 153, 174
    positive 71–4, 96–7
Utah 94, 135, 137–8, 150, 169–74

violence 12, 38, 71, 75, 168, 174, 191–7
voting 19, 23, 77–9, 80, 84, 86, 93

water markets 10–11, 161–74
water-diamond paradox 31–2
wilderness 40, 114–15, 149–53, 159

Yandle, Bruce 82–3, 104–6
Yellowstone National Park 44–5, 114–16, 149–52, 153

zero-sum game 8, 60, 80, 101, 103, 106, 159, 174, 193

ABOUT THE AUTHOR

Dr. Jordan K. Lofthouse is a Senior Fellow with the F. A. Hayek Program for Advanced Study in Philosophy, Politics, and Economics. He is also a Program Director of Academic & Student Programs at the Mercatus Center at George Mason University. He received a PhD in economics from George Mason University in 2020. Prior to his doctoral program, he received an MS in economics in 2016 and a BS in geography in 2014, both from Utah State University. His research centers on the Austrian, Virginia, and Bloomington schools of political economy. Some of his research topics include environmental economics, Native American economic development, entrepreneurship, and disaster recovery.